2,500 Years of European Helmets

2,500 Years of European Helmets

800 B.C.-1700 A.D.

Howard M. Curtis

BEINFELD PUBLISHING, INC.
North Hollywood, California

Library of Congress
Catalog Card Number 76-20423
ISBN 0-917714-06-7

First Edition

Published by
BEINFELD PUBLISHING, INC.
North Hollywood, California

Printed by
TAYLOR PUBLISHING COMPANY
Dallas, Texas

designed by L. S. Beinfeld

Table of Contents

PREFACE

The fighting man has always given priority to the protection of his head when choosing defensive equipment, and except in the most primitive societies, has usually worn a helmet even when for technical or economic reasons he has been unable to obtain any other body armour. The earliest recorded examples of armour are the copper helmets dating from about 2,600 B.C., found at Ur in Sumeria, and helmets have survived until the present day as the only kind of personal armour in general use in the armies of the world. A helmet is a complete object in itself, even though it may have been designed originally to go with a full harness, and so has a unity of form that makes it more attractive to collectors than other detached pieces of armour. Moreover, both because they were produced in larger quantities and because of their homogeneous construction, more helmets survive than any other types or armour, and for some periods are virtually the only type that does survive. Even the wealthiest collector can have little hope of obtaining a 14th-century breastplate, but he might, with luck, buy a 14th-century helmet; while at the other end of the scale the most impoverished collector would be very unlucky if he could not acquire at least a 17th-century morion or a trooper's pot. The helmet, therefore, holds a special place in the history of armour, a fact that has been recognised since 1878, when the first book devoted entirely to it, Gustav von Suttner's DER HELM, was published. This has been followed by many other books on the same subject though, rather surprisingly, the last general one in English appears to be the famous catalogue of the Exhibition of Helmets and Mail held in London by the Royal Archaeological Institute in June 1880. The publication of a new book in English is obviously long overdue, and it is to be hoped, therefore, that the present work by my friend Howard Curtis, with its large **corpus** of illustrations from many sources, will be useful to both collectors and students.

C. Blair
Victoria and Albert Museum
London S.W.

INTRODUCTION

Some thirty-five years have gone by since I first became fascinated with arms and armor, and for the past seven of these I have been planning this picture book on European helmets — the kind of book a collector, student, or just the curious would enjoy picking up and browsing through, either for reference or relaxation.

The helmet is the most sought after, in my opinion the most handsome and the most romantic of all the various components of a harness.

I have included as many examples as I could justify from the early plate armor period (c. 1330-1410); and the great period, so-called Gothic (1410-1500), not only because of their rarity but because of their great beauty of line, symmetry, and simplicity — helmets of the finest compositional quality from the iron ore deposits of Styria or the mountains of Northern Italy, forged by the hands of a few master craftsmen, closely guarding their precious secret process, men yet unaware of the percentage formula of carbon and iron resulting in a "sure fire" product of the hardest, most compact and durable steel; yet their products represented in one piece construction the epitome of the armorer's skill, a talent rarely developed today. It is no wonder these men were sought after by kings and princes, invited to their courts, and bestowed with gifts, special privileges and high salaries.

Their head-pieces varied in thickness, the crown and forehead requiring more protection than the back and brim against the lance, sword, axe, mace, and halberd. They stressed function and practicality, protecting the head from various specific types of exposure, and are works of fine art in their own right, as are indeed the less practical, highly decorative parade and show pieces of the 16th and 17th centuries.

My objective of encompassing all the various types of head-piece has suffered only through an equal emphasis on the above-mentioned "great period" helmets; and yet examples for the average soldier (armory issue, rather than expensive, tailor made products) have been included, some of which are rarities in museums and private collections today because they were once so common - therefore, later neglected, discarded, or melted down. A similiar tragedy occurred with the famed English longbow of the 14th and 15th centuries (which played such a significant role against the French in the Hundred Years' War) - a weapon once so vital and yet so common that it eventually succumbed, broken into pieces, to a wintry eve's warming fire. Another situation, which has contributed to the scarcity of homogeneous Italian helmets of the 15th century, has been the modification of a number of these masterpieces for use in the popular Pisan Tournament, the Giuoco del Ponte, composed of citizens of the north bank of the Arno against those of the south bank fighting over possession of the central bridge. The contestants used helmets on which original visors were replaced with 17th and 18th century barred face guards. Decorative wooden paddle-shaped shields called ''targones'' were used as offensive weapons as well, and were held with two hands so the combatant could both thrust and parry. These mock battles, fought with a club and a shield as early as Roman times, continued until the year 1807.

Construction of a helmet began with a single billet which was hammered into a plate and then worked in a cold state with frequent annealing to forge the bowl. The outer surface could not be hammered with the heavy blows used on the inside without sacrificing the thickness needed for protection when later the finished bowl was smoothed out. Then the sides were developed, the cranium modeled, the sides cut out for the fitting of cheek-plates if required, the visor and upper bevor (the term ventail is outmoded) were fitted if called for; and the helmet was finished by filing, polishing, piercing, engraving, etching, fire gilding or painting as desired.

The captions are arranged in chronological order, presenting an evolution of the European helmet through the 17th century. Marks, where evident, are described so they can be identified, but it has not been the purpose of this book to offer coverage of marks — a subject covered in part by a number of other volumes on armor in general, and in preparation more comprehensively by other writers. It is not my intent to assume any authoritative status. I have merely attempted to present a colorful, enticing and, if you will, exciting array of helmets, utilizing the superior knowledge of experts and the resources of some outstanding museums.

I want to express my appreciation to Mr. H. Russell Robinson, Assistant to the Master of the Armouries at the Tower of London, who donated much of his time on my visits to London; to the British Museum, London; to Mrs. Patricia R. Atkins of the Royal Scottish Museum, Edinburgh, for her assistance and for placing in my hands for a moment one of the helmets I have most admired, the 14th century jousting helm of Sir Richard Pembridge; to the Glasgow Art Gallery and Museum, Kelvingrove, Glasgow; to Dr. Bruno Thomas, Director, and Dr. Ortwin Gamber, Associate Curator of the Kunsthistorisches Museum, Vienna, for their cooperation on my visit and helpful correspondence; to Dr. Ernst Koniger, Curator of the Arms Department, Nuremberg Germanisches Nationalmuseum; to Prof. Dr.

Wolfgang Herbst, Director, Museum fur Deutsche Geschichte, Berlin; to Dr. Hugo Schneider, Vice Director of Arms and Military Material, Schweizerisches Landemuseum, Zurich; to Albert J. Gagne, Curator, John Woodman Higgins Armory, Worcester, Massachusetts; to Dr. Helmut Nickel, Curator of Arms and Armor, Metropolitan Museum of Art, New York and to Mr. Stephen V. Grancsay, Curator Emeritus, for his enlightening comments and wealth of knowledge; to the Los Angeles County Museum of Natural History and the Los Angeles County Museum of Art; to the Tojhusmuseet and the Danish National Museum, Copenhagen; to the Musee de l'Armee, Paris; to the Palazzo Ducale, Venice; to Marcello Terenzi, Curator of the Arms Collections at Castel S. Angelo, Rome; to my friend, the late Harold L. Peterson, former Chief Curator, National Park Service Museums and most of all to Mr. Claude Blair, Keeper, Department of Metalwork, Victoria and Albert Museum, London, for his kindness and patience in reading the manuscript, and without whose invaluable corrections and suggestions the manuscript would have suffered considerably.

I must apoligize for the few instances in which the writer has digressed from the subject matter and has shown a complete harness or contemporary object in an effort to provide a slight change of pace for the eye of the viewer; also despite the writer's appreciation for uniformity I have included weights and dimensions on some helmets where information was available and made significant contribution, yet omitted it on those where it wasn't available, feeling that it is better to provide some useful supplementary information than none at all.

I sincerely hope the reader will derive some of the pleasures from this volume that I have had in photographing and preparing it.

Howard Curtis
September, 1977

The Helmets

HORNED HELMET, VIKSBOG, NORTH ZEALAND, DENMARK, c. 800-400 B.C., Danish National Museum, Copenhagen.

This helmet features beautiful workmanship in bronze and was one of two found when cutting peat in 1942. The skull is cast from two pieces, their upper edges left jagged. The crest was cast separately and riveted into place over these edges. A slot in the middle of the crest, lengthwise, allowed a plume to be fastened with an adhesive; and short cruciform projections at the apex, obscured by the horns, have small sockets for the attachment of other ornaments. The front of the helmet represents a face, the eyes and eyebrows detachable. The small row of bosses around the rim terminates in birds' heads, two confronting each other at the front and back, apparently representing boats with a bird's head on the prow and stern. Two holes on each side near the rim served for attaching a lining and chin straps (or cheek pieces). The idea of wearing horns was very widespread in the Northern Bronze Age. Although there are representations of horned helmets on Danish bronze razors and on some Swedish rock carvings, this, and its companion, are the only horned helmets surviving from the Scandanavian countries.

HELMET, GREEK (CORINTHIAN), 700-400 B.C., Kunsthistorisches Museum, Vienna.

A lovely example of the famous bronze helmets of antiquity which were ''recalled'' to style by the Italian iron barbutas almost 2,000 years later. Note the holes for the lining.

BRONZE HELMET, GREEK, (ATTICA), c. 500-200 B.C.

The Athenians used Corinthian type helmets until the sixth century B.C. when their original ''attic'' helmet was developed, following a trend toward a headpiece of less weight. It exposed the ears, cheeks, and part of the forehead, but occasionally had cheek pieces and frequently, a neck guard. The hoplite Aristion wore a diminutive type, essentially a skull cap. This Athenian coin, a silver tetradrachm dating 188-187 B.C., shows the goddess Athena wearing a typical crested ''Attic'' helmet, of which few examples survive.

BRONZE HELMET, ROMAN, (MONTEFORTINO TYPE A), 4th-3rd C. B.C. Los Angeles County Museum of Art, William Randolph Hearst Collection.

This plain helmet, cast in mammiform style, is surmounted by a gadrooned knob and has a finely molded rim. Fortunately, the cheek pieces, ornamented with three rosette bosses, were found with it when it was excavated near Perugia, Italy, at the end of the 19th century. The peak-like projection is a guard for the nape of the neck, the helmet resembling a 'jockey cap' worn from back to front.

THE SUTTON-HOO HELMET, VENDELIAN, (SWEDISH), c. 5th to 6th CENTURY. The British Museum.

Of iron, originally tinned and partly gilded, it was placed in a memorial ship to an East Anglian king; found on the east bank of the River Deben, in Suffolk. The ship dates from c. 650, but the helmet is believed earlier, having been brought from Sweden by the ancestors of the king who inherited it.

LEGIONARY HELMET, ROMAN, c. 50 A.D.
The Chrylser Museum of Norfolk,
Gift of Cornelius R. Love.

Iron, Imperial Gallic Type, ear pieces and crest missing. Note the similarity of form with the seventeenth century lobster-tailed pot helmet. This measures 9½ inches by 9½ inches.

GLADIATOR'S HELMET, ROMAN, c. 300 A.D.
The John Woodman Higgins Armory, Worcester, Massachusetts.

The crest, bowl and brim are cast separately of bronze and riveted together. The hammered bronze bowl is of one piece decorated with two rows of bossets with a warrior's head between. A lion's head, also chased and cast, is at the top of the tall crest. The wide brim is of one piece and is riveted to the bowl by eight bronze bands. Overall weight, 5 lbs. 4 oz. Gladiators were first used at Etruscan funerals where the conquered man ''accompanied'' the dead as an armed attendant in the next world. At Rome, the first exhibitions increased in size from three pairs in 264 B.C. (at the funeral of a Brutus) to 300 pairs in Julius Caesar's time, and extended from one day to as many as a hundred under Titus' reign. Trajan had 500 pairs. Amphitheaters were built in other towns of the empire. Gladiatorial combat was abolished by Constantine I in 325, and again by Honorius (393-423). It may have sprung up a few more times before it disappeared completely.

CONICAL HELMET (SPANGENHELM), OF A BARBARIAN CHIEFTAIN, FRANKISH, 6th CENTURY. Metropolitan Museum of Art, New York.

Four iron segments within a bronze framework, chased with foliation and birds, at the base of which are seen holes for the attachment of the lining; only a trace of the nasal guard remains.

CONICAL HELMET, POLISH, 9th — 10th CENTURIES.
Tower of London Armouries, (Loan from the Liverpool Museum)

Iron, covered with gilded copper, the four segments fastened with silver rivets. This type of helmet is called in modern Germany a Spangenhelm; that is, a round or conical head-piece often divided into sections by broad bands of iron springing upwards from a horizontal band around the lower rim. This example does not appear to have had any reinforcing framework.

CONICAL HELMET, (NASALHELM), MÄHRISCH (FOUND IN MORAVIA), 11th CENTURY. Kunsthistorisches Museum, Vienna (WSA41).

Forged entirely of one piece of iron, this helmet shows the type of construction used on some of the finest products of the most gifted armorers of the period. The same type, with variations, was used at the Battle of Hastings, 1066. This one is possibly as early as the 9th or 10th century. Similar helmets may be seen in the Maciejowski Bible, c. 1250, in the Pierpont Morgan Library.

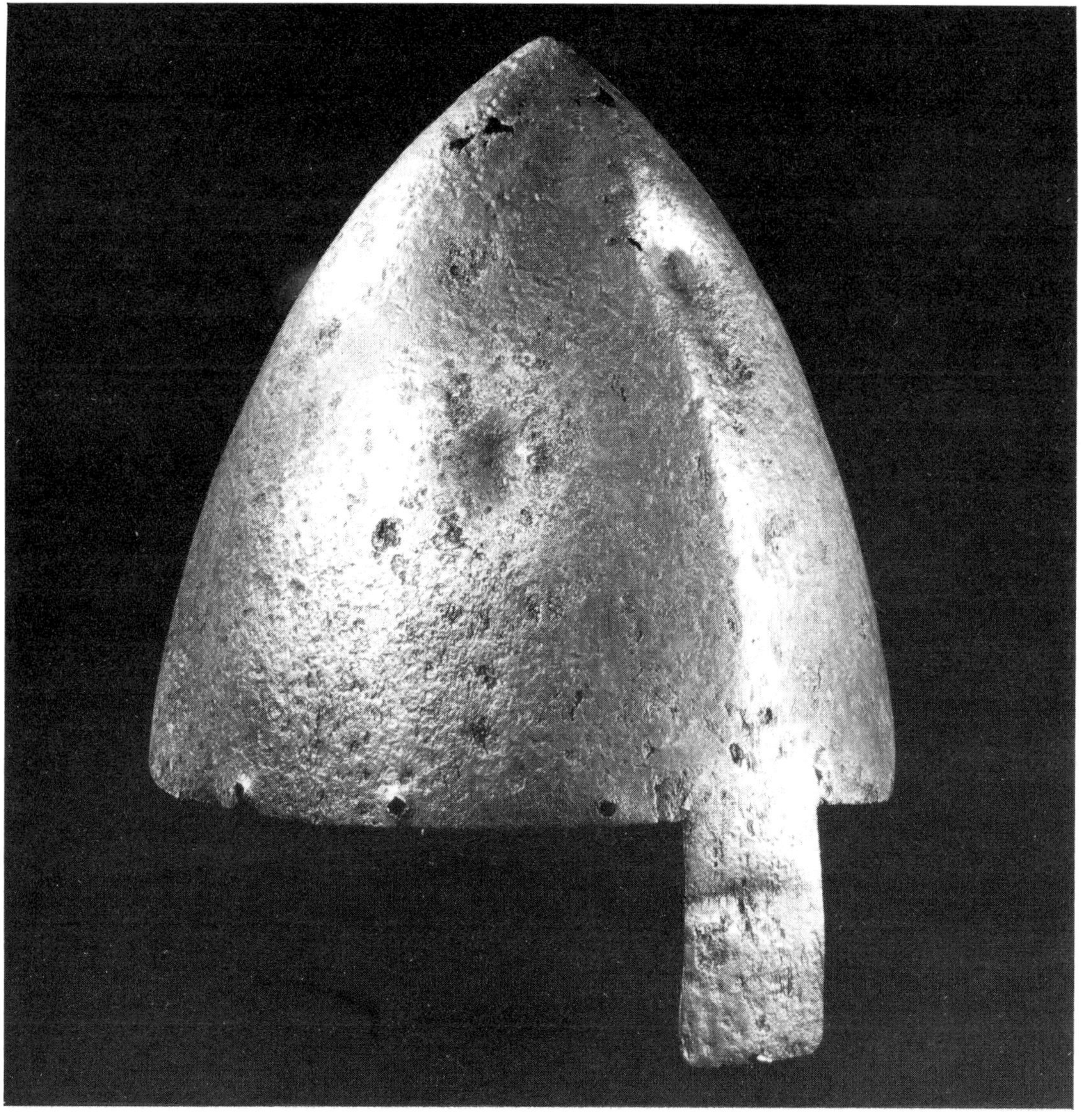

CONICAL, HELMET, NORMAN, 12th CENTURY.
The Metropolitan Museum of Art, Bashford Dean Memorial Collection, purchase funds from various donors, 1929.

The cheek pieces prevalent on similarly constructed helmets of the Roman and Migration periods were not in use at this time. This Spangenhelm is constructed of segments overlapping vertically and riveted together, without any framework of iron bands. The nasal guard and the frontal skull segment are forged as a single plate of iron. These helmets were usually worn over a mail coif. This one measures 7½ inches high, the segments (four) are 6½ inches wide at the base; those of the front and back overlap the side panels. The nasal piece is restored. It was found in northern France.

GREAT HELM, (POT HELM, TOPFHELM), GERMAN, 2nd HALF OF 13th CENTURY, Museum fur Deutsche Geschichte, Berlin.

This lovely battle helmet was excavated from Deerberg at Bubach (Schlossberg Dargen) in the Province of Pomerania. It is constructed of five riveted iron plates; two forming the bottom half, two forming the upper half, and the flat crown or deckplate. It has a vertical reinforcing band in front as well as horizontal bands above and below the eye slits. Both lower front sides are pierced with ventilation holes. 5 lb. 10½ oz.; front height, 11½'', back height, 10''.

GREAT HELM, (POT HELM, TOPFHELM), GERMAN, 2nd HALF OF 13th CENTURY, Museum fur Deutsche Geschichte, Berlin.

Side view.

"The Siege of Antioch," illustrations from a 13th century history of the crusades. **Histoire de Jerusalem,** Ms. Fr. 9081, fol. 44, Osterreichische Nationalbibliotek, Vienna.

Three types of head protection of the 13th century are illustrated here. The Conical and round topped helms with nasals, top, center and middle right, are a carry over from the eleventh century. Individual mail coifs were worn as early as the eleventh century also. The Knights at upper right all wear coifs extended from their hauberks in one piece. A small steel skull-cap (cervelliere or bascinet) was sometimes worn beneath the mail coif, as was an arming cap which laced under the chin and had a padded ring which cushioned the skull. The arming cap was worn over the mail coif when the third type of helmet was used, the cylindrical flat-topped helm seen at left and center in the painting. Towards the end of the century the skull of these helms became more pointed to deflect downward sword cuts, and evolved into the great helm of the fourteenth century.

GREAT HELM, (POT HELM, TOPFHELM), POSSIBLY SWISS, c. 1300, Kanton Museum Basel-Land, Liestal.

This is one of two pot helmets excavated from Madeln Castle on Adler mountain in Basel-Land County, Switzerland, which was damaged by an earthquake in 1356. It consists of five riveted iron plates, two each in the top and bottom half and the deck plate. The small band (nose guard) between the 5'' eye slits is riveted to the upper front plate. There is a slight median ridge on both the front and back. The dent just below the crown on the right side was the result of earthquake damage. Thirty-eight round airholes are pierced on the lower right side for ventilation; and the deck plate, riveted to the inward bent rim of the upper surface, has 2 holes for helmet ornament straps. The pierced cross on the lower right side was for a toggle bolt which had a retainer chain attached to the breast. Because it was found in such badly rusted condition, it was necessary to separate the plates, treat them properly to prevent further oxidation, and reassemble them, restoring small areas as necessary. The height at the side is 10¾''; the deck plate measured 7½x6-3/8'', and the existing weight is 5½ lbs. (some of the original weight has diminished through corrosion). This represents the earliest preserved helm of this type, with the exception of the preceding one at the Museum fur Deutsche Geschichte in Berlin. These are distinguished by their flat deck plates and construction of five plates (see next helm).

GREAT HELM, (POT HELM, TOPFHELM), PROBABLY SWEDISH, c. 1300, Statens Historiska Museum, Stockholm, Sweden.

The Construction consists of three riveted iron plates. There are two plates composing the lower section, the front one pierced with 11 T-shaped holes, (air vents) on each side, below the visor. The third plate, forming the skull piece is flattened on top. A narrow projection, continuing upwards from the central keel of the lower section serves as an eye slit divider and forehead reinforce. This lovely helmet was found at Burg Aranas, West Gotland, and is an example of a helmet designed purely for battle. Gotland is an island in the Baltic Sea off the south-east coast of Sweden, the site of the Battle of Wisby, July 27, 1361, in which the Danes attacked the Swedish town.

GREAT HELM, (POTHELM, TOPFHELM),
FOUND AT BOLZANO, ITALY,
PROBABLY GERMAN, c. 1300,
Castel Sant' Angelo, Rome.

The pairs of holes pierced below the top plate are for mounting a crest with laces. This helmet did not rest on the shoulders and was leather lined and strapped. It sat on top of a linen cap or iron skull cap beneath which was folded the long hair of the warrior to act as a cushion. It retains the flat top predominant on large helmets of the late 12th and early 13th century, used for war.

BASCINET, (BASCINETTO), ITALIAN, c. 1350, The Metropolitan Museum of Art, New York.

This 6 lb. helmet is forged from one billet of steel, the bowl measuring 14 inches in height. It came from Chalcis, the principal town of the Island of Euboea, Greece; the scene of many medieval battles between the Turks and Venetians.

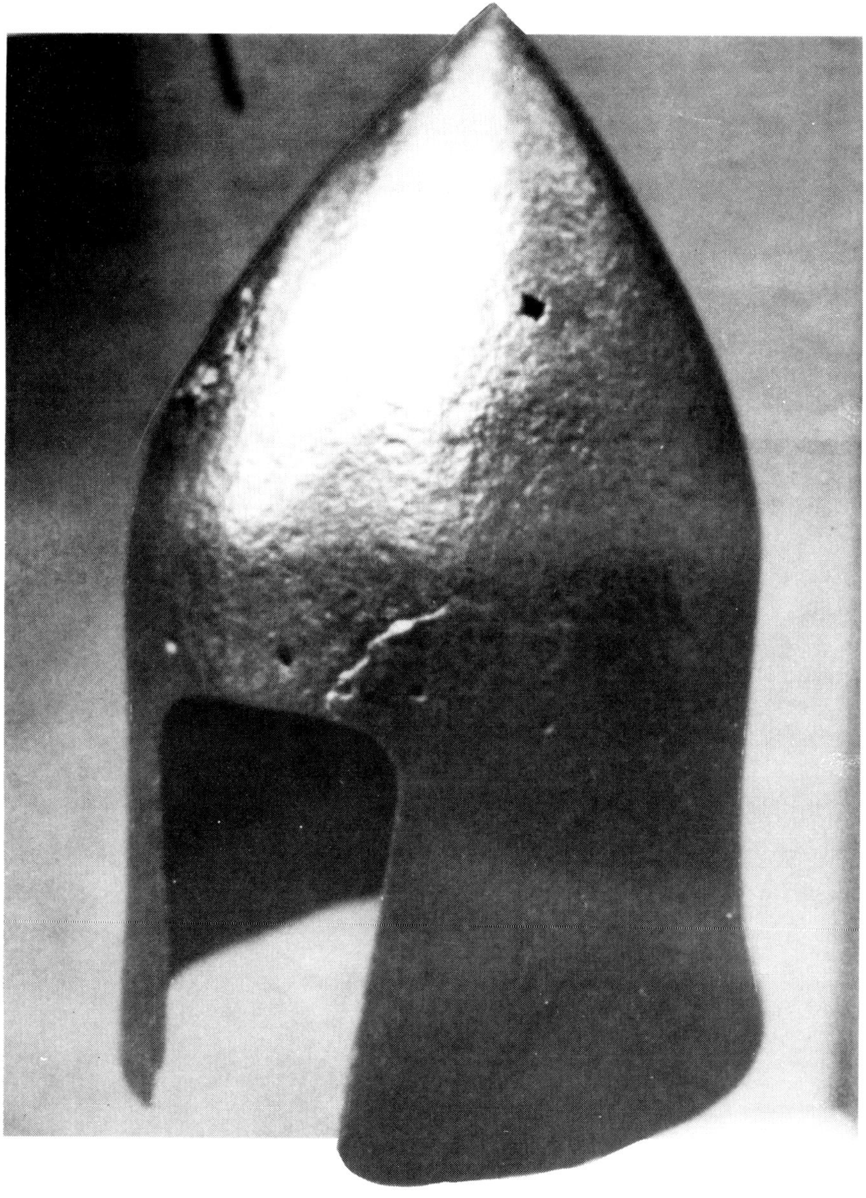

BASCINET, (BECKENHAUBE), GERMAN, 1300-20, Glasgow Art Gallery and Museum, Kelvingrove, Glasgow.

Found at Braunfels Castle, near Frankfurt. The two pierced holes at the sharp bottom angle undoubtedly secured a chin strap. The other holes were for rivets attaching a lining, and an aventail to protect the face, neck and chin. In the 14th century the pointed skull replaced the hemispherical shape on bascinets. After about 1320, staples (vervelles) were used for attaching the lacing rings at the upper border of the aventail to the helmet.

GREAT HELM, (POT HELM, TOPFHELM), POSSIBLY SWISS, c. 1325-1350, Schweizerisches Landesmuseum, Zurich.

This helm was found in 1917 at Kussnacht Castle in Switzerland (Kanton Schwyz), and was constructed of five plates secured by four rivets at each joint. The hemispherical deck plate has diminished in size from the earlier types (see the Madeln and Bolzano helms) due to the stronger slope (conical shape) of the skull section. The eye slits offer more protection with the development of the outward turned flanges above and below. Its existing weight is approximately four pounds.

GREAT HELM, (POT HELM, TOPFHELM), GERMAN, c. 1350.
Hessiches Landesmuseum, Darmstadt, West Germany.

This helmet is constructed from five separate riveted steel plates, two each composing the lower and upper sections; the crown an oval semi-hemispherical plate. The upper surface of each eye slit is bevelled slightly forward serving as a deflector from downward blows of a sword. Numerous ventilation holes are pierced on the right side, as is a large cross for the attachment of a toggle on a retaining chain anchored to the breast. This helmet was intended for use for either the tournament or war, and was found at Burg Tannenberg, Hesse, which lies between the Rhine plateau and the Thüringer Wald.

GREAT HELM, (POT HELM TOPFHELM), GERMAN, c. 1350, Germanisches Nationalmuseum, Nuremberg.

This was part of the funeral achievement of one of the Rieter family from Kornburg Castle, Weihern, found in the Allerheiligen bei Kleinschwarzenlohe Church. It is a beautiful specimen, retaining an original mail tippet (aventail, camail) to protect the neck and painted decoration. Tippets were in use in the late 13th century, but were rare before 1320. This helm exemplifies a transitional step in that it is constructed of three riveted plates rather than five. The lower section is composed of two plates (front and back). The upper half is of one plate with a strong vaulted crown piece. Weight, 6½ lbs.

GREAT HELM, (POT HELM, TOPFHELM), AUSTRIAN SECOND HALF OF 14th CENTURY.
Oberosterreichisches Landesmuseum Linz, Scholssmuseum, No. c 1765.

This helmet unfortunately suffered a great deal of corrosive wear in the River Traun, near Linz, where it was found, and yet is still a lovely and desirable helmet. It is constructed of three principal plates of steel, the lower ventilation holes, on both sides of the lower-front plate, are of a crown of one piece. A steel band riveted around the slightly arched top undoubtedly served as a support for a large crest, anchored from the small central mount. The rivets around the band also secured the upper lining. The ventilation holes, on both sides of the lower plate, are of a long rectangular, small square, and small round shape.

GREAT HELM, ENGLISH, THIRD QUARTER OF THE 14th CENTURY, Formerly Peter Dale Collection, now in the Tower of London Armouries.

This lovely head piece is the most recently "prominent" of 15 great helms known to exist today, the third one found in England and of English manufacture. It is of identical construction to the Sir Richard Pembridge helm, and very likely from the same workshop. It has been "hidden" in private collections and only recently come to light. It is not yet known to whom it belonged but is the sort of helm which would have been worn by a knight in the army of the Black Prince in battles such as Crecy (1346) Poitiers (1356) and Najera (1367). It has been suggested it was the helm of Sir John Chandos.

GREAT HELM, ENGLISH, THIRD QUARTER OF 14th CENTURY, Royal Scottish Museum, Edinburgh.

This lovely helmet belonged to Sir Richard Pembridge, died 1375, an early recipient of the Order of the Garter. It stood on a perch over his effigy in Hereford Cathedral until it was presented to Sir Samuel Meyrick in 1882. During the last half of the 14th century this type of helm was relegated to tournament use primarily. It was worn over the open bascinet, or carried slung over the back when not in use, by a chain and toggle fitting into the cross shaped hole visible at the lower front. Composed of three pieces: a cylinder, a conical piece and a top piece secured with 3/16'' diameter nails with hemispherical heads turned over to the inside. The welds on the two lower pieces are so skillfully done they do not show. Every third nail on the level of the occularium slit has a diamond shaped washer for mounting the leather strip to which the lining was sewn. Traces of original leather exist around the internal rivets. The edges of metal turned outwards around the eye slits are very thick, protecting the eyes. The bottom edge of the helmet is rolled inwards over a thick wire so the surcoat won't be cut. Pairs of holes around the top are for attachment of the crest with lacings. Weight 5 lbs. 12 oz. Originally this helmet was burnished in a beautiful glossy black color.

GREAT HELM, QUILTED SURCOAT, SHIELD, AND COPPER GILT GAUNTLETS OF THE BLACK PRINCE, ENGLISH, THIRD QUARTER OF THE 14th CENTURY. Canterbury Cathedral.

The helmet retains its original parchment leopard crest. Its crown is now missing as is the Prince's silver label about its neck. The leather cap upon which the leopard stood was originally painted red and lined with red velvet. The crest weighs 4 lbs. 5 oz. and was entirely gilt. The helmet was worn over the cervellière (small skull cap bascinet) and coif of mail. A leather lining was fastened in the crown by rivets below the apex. The leather was cut into a series of triangular gussets adjustable with a cord, much the same as our early American football helmets. Part of the guard chain remains, which attached the helmet to the breast plate or coat of plates by means of a bolt. It passed through one of the pierced crosses in the lower front of the helmet. The helmet is 14 inches high and weighs 7 lbs. 2 oz. and consists of three pieces. front and back, welded together at the sides, and a thicker conical top piece. The right side of the helm is pierced with 88 ventilation holes arranged in the form of an open crown. Every other rivet encircling the helm has a small washer for attaching the lining. Just below these are nine pairs of holes for the lining laces. Four pairs of holes over the top piece, from side to side, secured the crest. holes at the back were for securing the helm with a strap. The eldest son of King Edward III, the Black Prince, Prince of Wales, is noted for his great victory over the French in the One Hundred Years' War at Poitiers in 1356. He was one of the original members of the Order of the Garter, founded by his father c. 1348, consisting of 12 companions each for the King and the Prince, as at a tournament. Legend has it that the Order was founded as the result of an incident in which the King picked up a garter dropped by the Countess of Salisbury at a dance. When they laughed at him for this, the King rebuked them with the words ''Honi soit qui mal y pense'' (shame on him who thinks ill of it). A requirement for membership was knighthood, which the Prince of Wales won with his spurs at the Battle of Crécy, 1346, at age 16.

Sketches of Black Prince's Great Helm.
(See previous photo.)

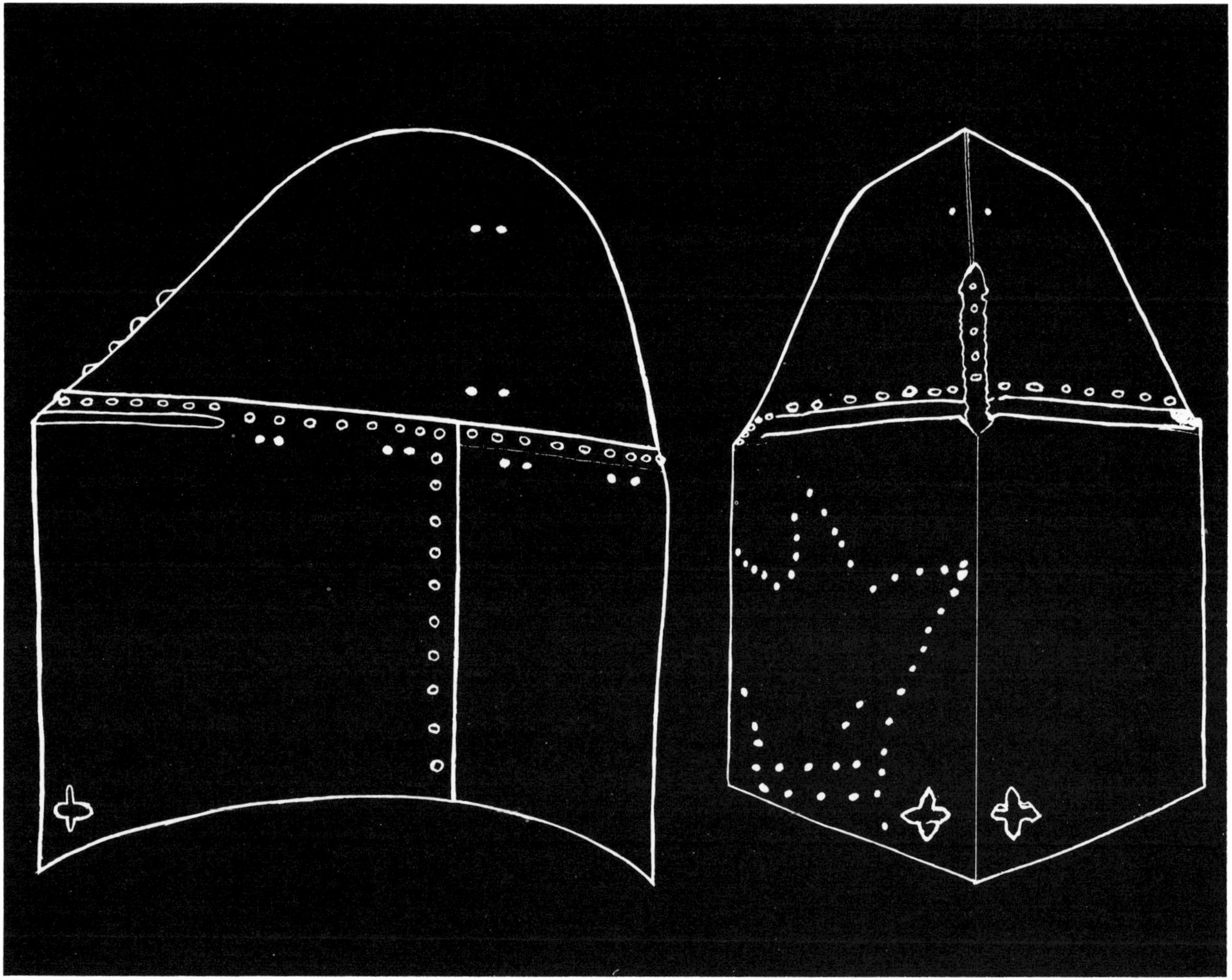

BASCINET (BECKENHAUBE), GERMAN, c. 1370.
The Metropolitan Museum of Art, New York.

Weighing 7¼ lbs., this has a visor of the Klappvisier type, a German innovation which was hinged at its top center to a vertical bar pierced to fit over two studs on the brow of the skull. (The cone at the apex of the skull is interesting, almost resembling a crest mount of antiquity.) Klappvisiers were most popular in Germany, but have been pictured in English and Italian paintings.

BASCINET, (BECKENHAUBE, BACINETTO), PROBABLY GERMAN, c. 1370.
Private Collection.

This retains half of the hinge for its klappvisier, and a portion of the mounting for its mail aventail (far side). Private Collection.

**BASCINET (BECKENHAUBE), GERMAN, c. 1380.
Tower of London Armouries (IV.467).**

Another fine example of the ''Klappvisier'' type. Note the vervelles around the border of the skull and the mouth slit for ventilation. These bascinets with pointed visors were called ''pig-faced'' bascinets by Victorian enthusiasts, and are still called ''hundsgugel'' (hound's hood) or ''houndskull'' bascinets by collectors today. In medieval England the correct term was ''visored'' bascinet.

BASCINET (BASCINETTO), ITALIAN, c. 1380.
The Metropolitan Museum of Art, New York.

This helmet weighs 6¼ lbs. The conical snout on the visor became popular about this time. Note the line of openings over the mouth; the vervelles for attachment of the aventail; the hinge pins at the side, often guarded with a safety chain, which when pulled, released the visor or allowed it to pivot to one side; and the strong backward slope at the front of the skull.

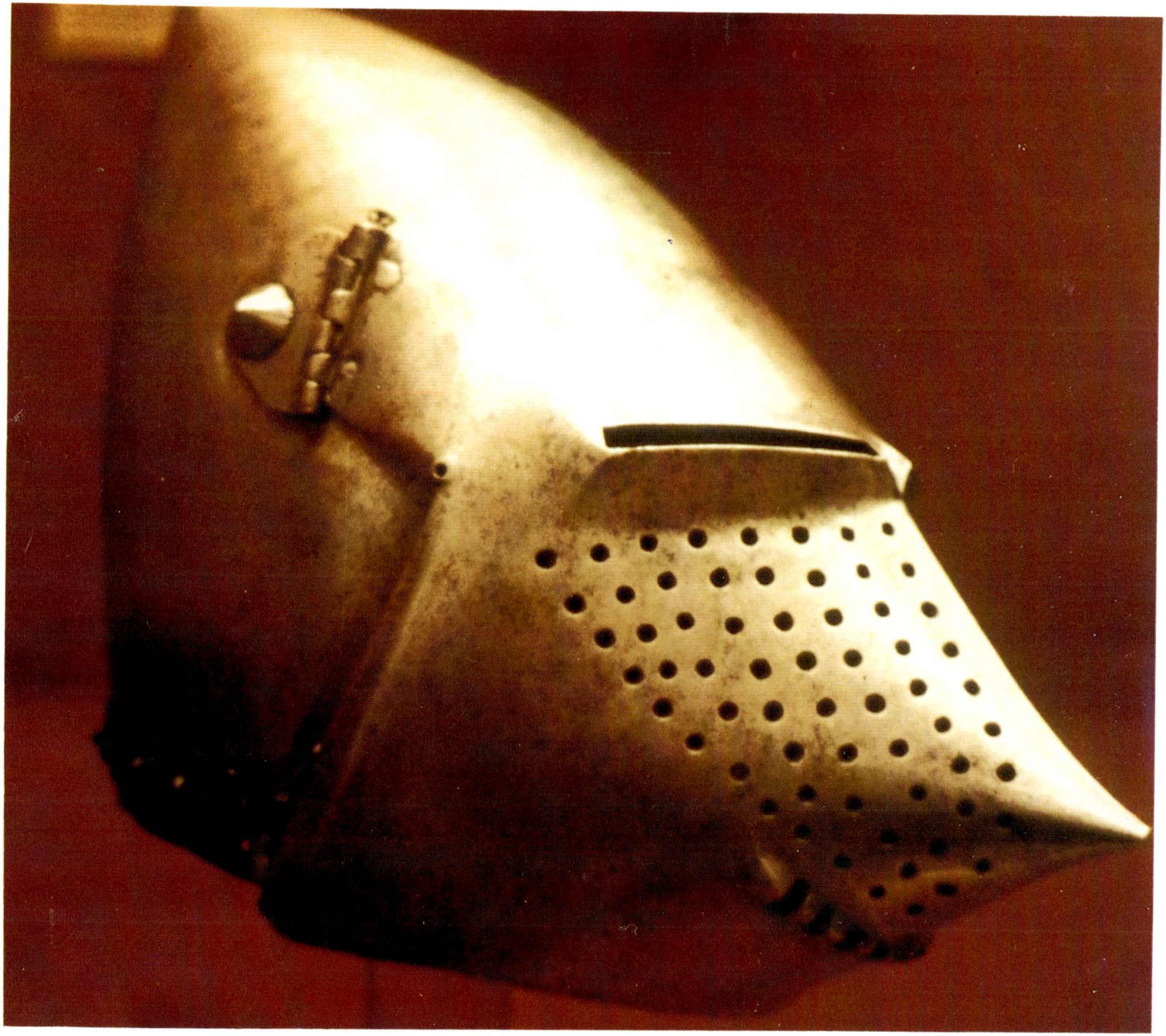

GREAT HELM ("THE LEBUS HELM", TOPFHELM TØNDEHUELM), PROBABLY GERMAN, END OF 14th CENTURY, Tøjhusmuseet, Copenhagen.

For war and joust, it was found at Lebus near Frankfurt an der Oder. Note the reinforcing plate for the joust visible at the front of the face. This helm appears to have been constructed of five plates originally, exclusive of the reinforcing plate. One plate at the lower back is missing. This type of helmet was worn over the mail coif and arming cap, had its own padded lining, and was secured with a chin strap. It evolved from the conical, flat topped, and round topped helmets, sometimes displaying face guards, pierced with eye slits and ventilation holes, used in the 12th and 13th centuries. Weight, less missing plate, 9.93 lbs.

GREAT HELM, (THE PRANCK FAMILY), GERMAN, END OF 14th CENTURY, Kunsthistorisches Museum, Vienna.

Displaying an original crest in the form of two curved horns, made of moulded silver and gilded leather and canvas, mounted on a leather cap attached to the skull by laces. It has a reinforcing plate designed to protect the left side of the face in the joust. Notice the taper to the top of the crown, sometimes evident after 1250. It hung in the Cathedral church of Seckau, in Steiermark, where it was suspended above the tomb of Pranker von Pranck; it then went into the famous Herr Franz Thill Collection.
Weight, 11.35 lbs.

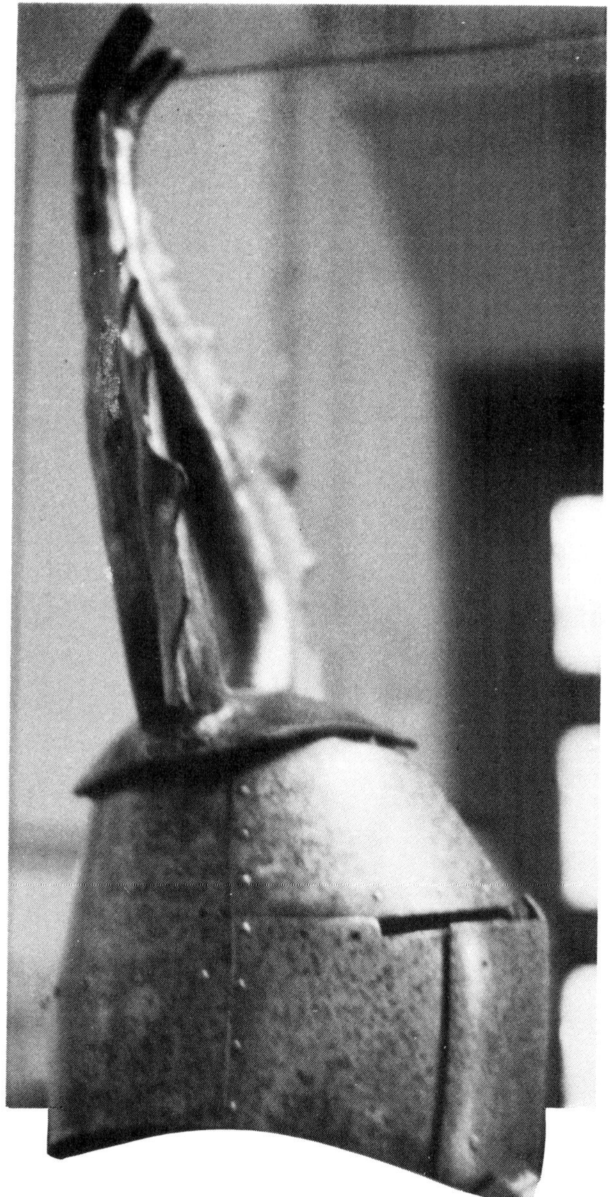

JOUSTING HELM, ENGLISH, c. 1380.
From the tomb of Sir Nicholas Hawberk, died 1407, Cobham Church, Kent.

This helmet is very similar to the helm of King Henry V in the Abbey Church of Westminister, but lacks the brass band added to the base. It is made from five metal plates, those at the bottom edge are rolled outwards. The four staples on the top of the skull are for attachment of a crest. The ring at the lower front, if it is original, was for attachment of a tie-down strap as a method of securing the helm to the breastplate. Permanent loan to the Tower of London Armouries. British Crown Copyright—reproduced with permission of the Controller of Her Brittannic Majesty's Stationery Office.

JOUSTING HELM, ENGLISH, c. 1380.
From The Tomb Of Sir Reginald Braybrook, Died 1405, Cobham Church, Kent.

After its original manufacture, the height of this helm was increased by addition of a metal band with a horizontal projection at the rear, probably for mounting about the tomb. The hole at the apex was for mounting of the crest. Note that on this helm, and that of Sir Nicholas Hawberk, there is no central division to the sight, the lower edge of which projects slightly forwards of the upper edge. These characteristics illustrate the transition from the 14th century great helm to the ''frog-mouthed'' tournament helm of the 15th century, q.v. Permanent loan to the Tower of London Armouries.

VISORED BASCINET (BASCINETTO, CON VISIERA CURVA), NORTHERN ITALIAN, c. 1380-1400.
Tower of London Armouries. (IV. 470).

One of the two finest in existence today, this one came from the famed armory at Churburg Castle. It is in superb condition, with decorative engraved brass borders and its original aventail. The vent holes omitted on the left side of the pointed visor indicate jousting was being done with the lance couched to the left over the horse's neck, opponents passing left side to left side. The smooth side of the helmet would deflect a lance more readily. The design of these helmets is such that a blow with a weapon from almost any direction, except a sidewards cut, would glance off harmlessly.

GREAT BASCINET, WITH BEVOR, (GRAND BACINET A METONNIERE), FRENCH, c. 1400, Musee de L'Armee, Paris.

This type was essentially a transition helmet between the smaller bascinet and the armet, distinguished by its rounded hinged and pivoted visor. The bevor and front gorget plate are separate pieces locked in place below the visor. A similar helmet, the bicoquet, was used in Germany, France, and in England, where it was called a ''bycocket.'' The bicoquet, if actually any different from the great bascinet, had a flattened hemispherical visor which conformed to the chin line, and sometimes had no visor slits, vision being adequate through many large ventilation holes. The chin and neck were protected by one or more large plates of iron, hinged at the top and fastened with a turning pin in front.

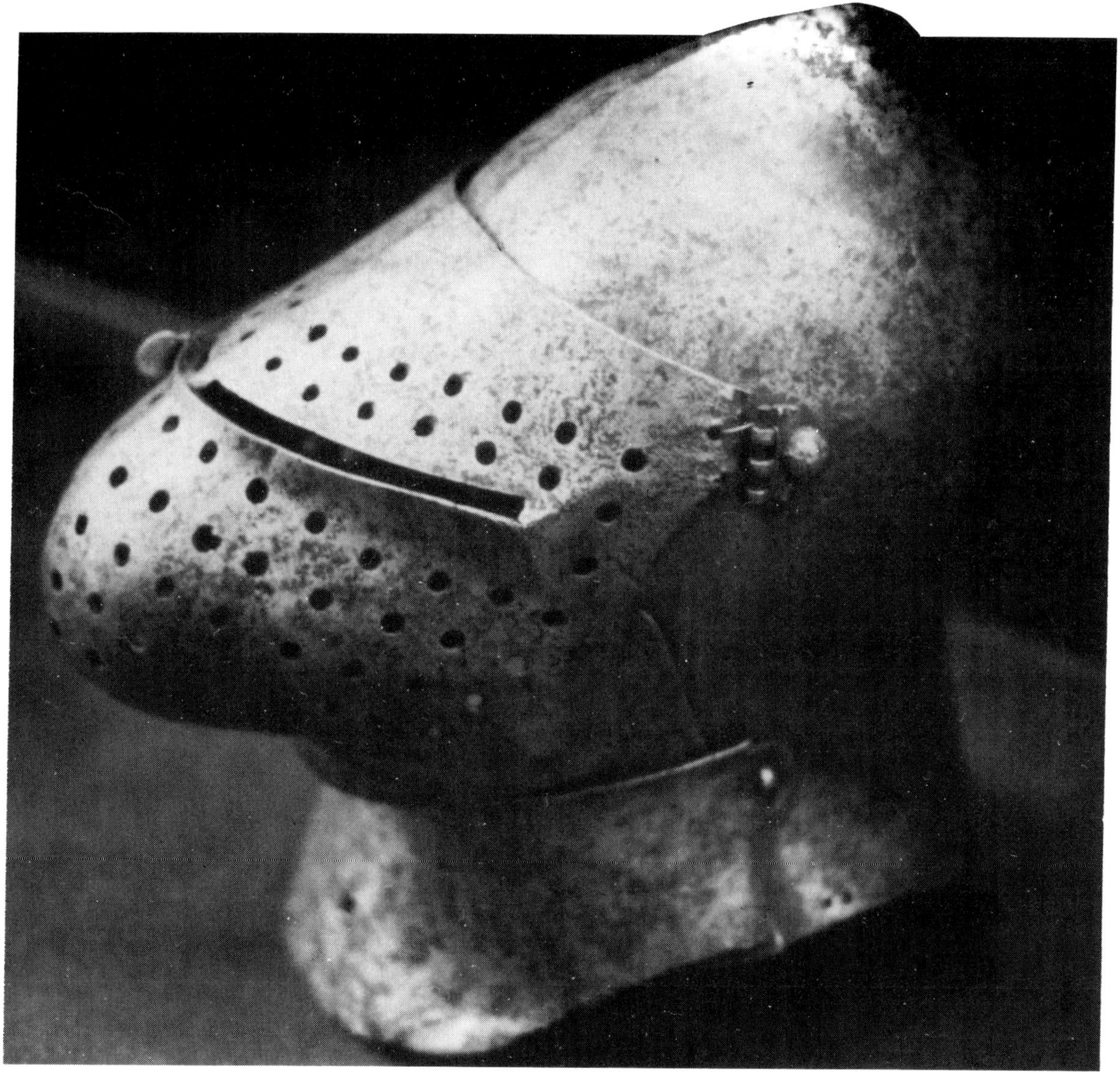

GREAT BASCINET, PROBABLY ITALIAN, c. 1400, Ducal Palace, Venice.

One of the most impressive medieval helmets I have ever seen, this one is forged entirely from one piece of iron (with the exception of the pointed visor), the solid plate bevor continuing as an extension of the skull in the shape of the mail aventail used with the smaller bascinets.

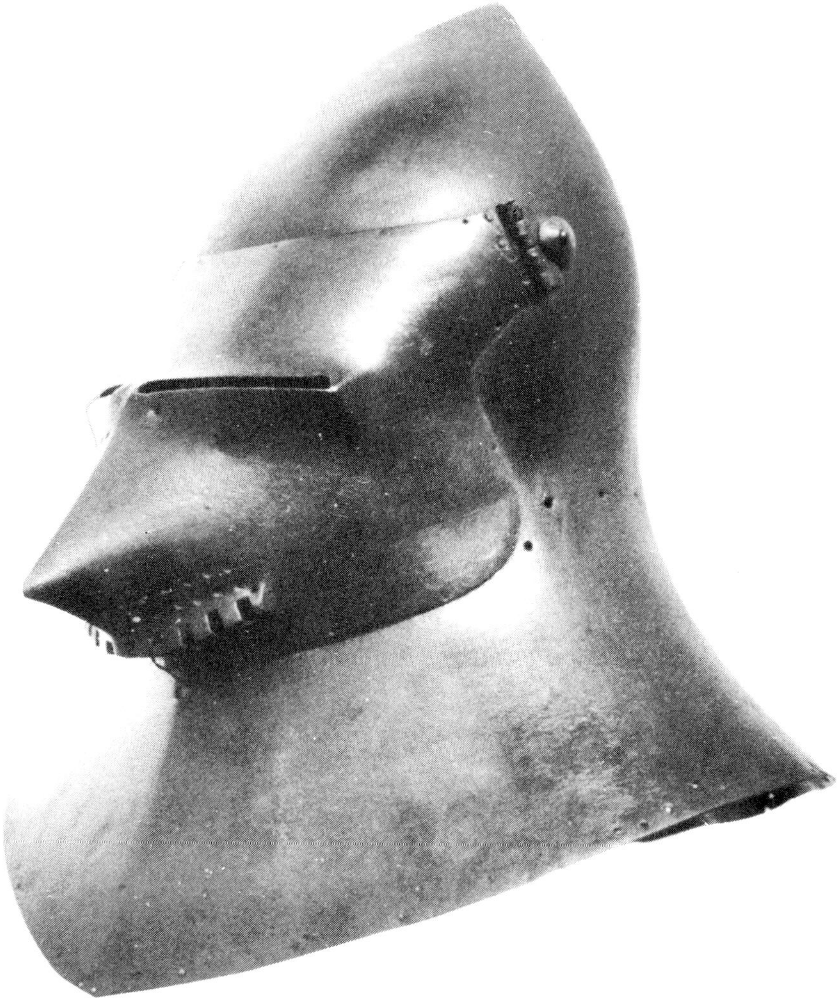

BASCINET (BACINET), FRENCH, c. 1400, Metropolitan Museum of Art, New York.

This was believed to have been one of Jeanne d'Arc's, although it is actually a bit too early. It hung above the main altar at the church of Saint Pierre de Martroi at Orleans. The chain remnant at the apex suggests it was suspended as an exvoto; however it may have been for a safety retainer, such as was used on Great helms when not in use. The two holes on the right side, center, were made by arrows or crossbow bolts. Weight, 5 lbs. The visor is missing.

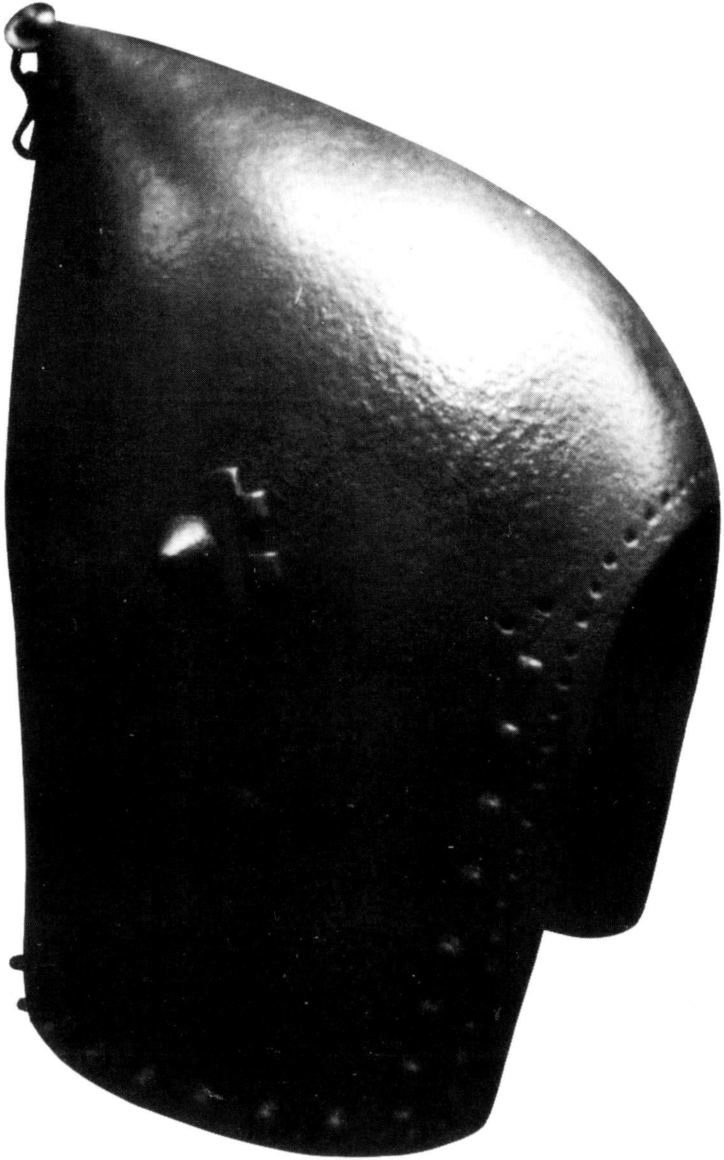

VISORED BASCINET,
(BECKENHAUBE MIT AUFSCHLAGBAREM SCHNAUZENVIZIER),
GERMAN, c. 1400,
The Metropolitan Museum of Art, New York (29.158.29)
on loan to the Los Angeles County Museum.

The conical bowl measures 9¾'' high and has a slight mid-ridge running halfway down the front and back. The apex is embossed and contains a brazen internal thread for securing a crest holder. Above the holes for the lining strap rivets, at the back and sides, are brass vervelles over which passes the leather band to which the aventail is wired. The aventail of mail is tailored to fit over shoulders, chest, and back, and comes to a point in front. The visor is associated, and has seven series of circular perforations in groups of three and four. There is an indentation between the vision slits above which are three more circular perforations. This German helmet has a style of visor perhaps more common in Italy at this period.

GREAT BASCINET (GRAND BACINET), FRENCH, c. 1400.
Musee de l'Armee, Paris.

Note the reinforcing plate across the chin area, which appears to guard the junction of the visor with the bevor. Turning pins lock the bevor in place, and the front gorget plate is detachable as on the example on page 75. This type of helmet was most popular for foot combat in the ring, and as one can see, gave a great deal of protection.

BASCINET (BASCINETTO), NORTH ITALIAN, 1380-1410.
Tower of London Armouries.

This beautiful styling, in addition to deflecting a downward blow, had the aesthetic appeal of the pointed arch, contemporarily popular in architecture. The visor is missing.

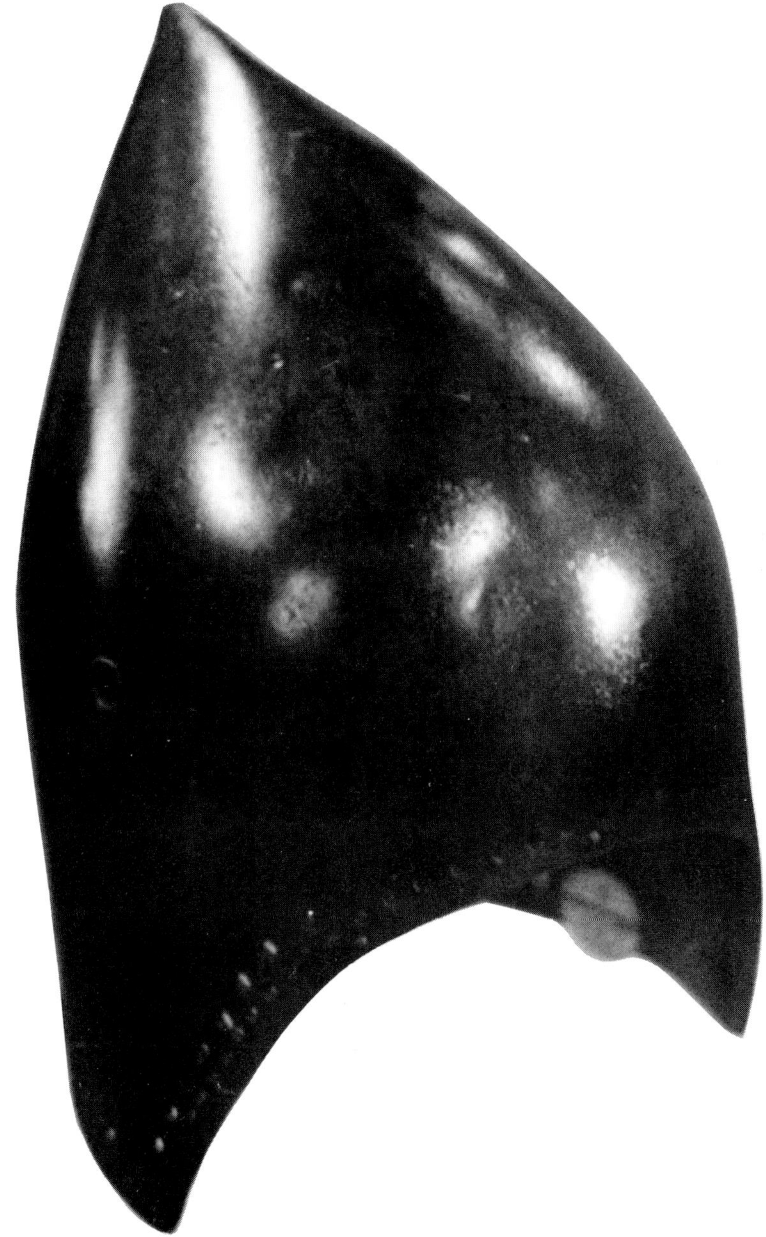

BASCINET, ITALIAN,
c. 1380-1425.
Author's Collection.

Measuring 12¾ inches in height. This unique helmet was made in two pieces, the skull and a cheek and neck defense. These are joined in a horizontal row of brass rivets above the eye level, several of which are missing. The visor is missing from the ''U'' shaped face opening. It is a marine excavation from off the southeast coast of Cyprus, and is somewhat similar to some of the early bascinets found at Chalcis Castle on the eastern Greek island of Euboea shown on Plate 52, figures 1-4, in Charles Ffoulkes article on Chalcis helmets at the Ethnological Museum in Athens, (**Archaeologia,** Volume 62, 1911). These latter helmets, however, appear to have been constucted of three or more pieces, and with separate (hinged) cheek defenses.

KETTLE OR WAR HAT,
(EISENHUT, CHAPEL-DE-FER, OR SHAPEWE IN
15th — 16th CENTURY ENGLAND); SOUTH GERMAN, c. 1430.
Tower of London Armouries.

This one displays a mail aventail. A very practical headpiece, the Kettle Hat was easily put on and taken off, and allowed great freedom of motion, while the broad brim protected the face and neck merely by moving the head in the direction desired.

BASCINET, (BACINETTO), ITALIAN, c. 1430.
The Metropolitan Museum of Art. (29.158.46).

This represents a final transition between the bascinet and the armet, the skull being pointed like earlier bascinets. It has unusually high cheek pieces; the right one is hinged at the side, opens in the front and overlaps the left. The vertical bar below the visor locks it and the cheek pieces with a turning bolt. From Chalcis, it weighs 8½ lbs.

SALLET, (SCHALLER, CELATA, SALADE), PROBABLY ITALIAN, POSSIBLY FRENCH OR FLEMISH, 1st HALF OF 15th CENTURY.
The Metropolitan Museum of Art, New York (04.3.277). Rogers Fund, 1904. Formerly Duc de Dino Collection.

Small and of rather unusual shape, this lovely helmet is in superb condition, and retains its original half visor. It was probably worn with a small bevor, and may well date as early as 1430. Rivets for the lining strap are missing. Weight 4 lbs. 2 oz. This helmet (and the one which follows) might also be designated ''Burgundian''; however, since Flanders was the industrial part of Burgundy, ''Flemish'' would be more accurate.

**SALLETS, GERMAN,
2nd QUARTER OF THE 15th CENTURY.**

Two combatants sit near the lists awaiting the eminent contest. Each has his stretcher beside and his attendant in front of him. They wear Gothic armor with open sallets and bevors (note the brow reinforce plates). From Hans Talhoffer's Schlichtegroll, Ambraser Codex, 1459. This volume concerns itself with the settling of disputes or judicial matters by means of a duel between the two persons involved.

Hie sitzent sie beid Im
schranken vnd wartent
des anlas vnd hat ÿeder sin
bär hinder im vnd sin grieß
warten vor im

SALLET, (SALADE, CELATA, SCHALLER), PROBABLY NORTH ITALIAN, POSSIBLY FRENCH OR FLEMISH, c. 1430-40. Tower of London Armouries.

This type has the visor slit in the hinged brow reinforce. These are comparatively light and maneuverable helmets, shaped in front so as to ward off a blow and with a pointed tail or neck guard in back, which was sometimes jointed in the German style. It was also a very popular helmet in 15th century France, and Jeanne d'Arc wears one on her head in the polychromed stone sculpture at the Musee Historique D'Orleans. The sallet was supplemented with the addition of a chin guard (bevor) fastened to the breastplate and strapped around the neck. It remained popular in Germany even after the armet replaced its use in England.

ARMET A' RONDELLE, (ELMETTO), ITALIAN, c. 1440.
Metropolitan Museum of Art, New York.

Twice inscribed by its maker, LIONARDO. Note the vervelles retained on early armets for attachment of a small aventail; the sparrow beak style of visor; and the fact that the hinges at the top of the cheek pieces (which open outwards when the visor is raised) are internal for protection. This helmet also has a full brow reinforce plate, barely visible in the photograph. The decorative rondel at the base of the neck anchored straps to an optional reinforcing bevor. The bevor also prevented the visor from being forced up. Weight, 9¼ lbs. The sparrow beak visor was not as popular in Germany after c. 1510 as it was elsewhere until c. 1530.

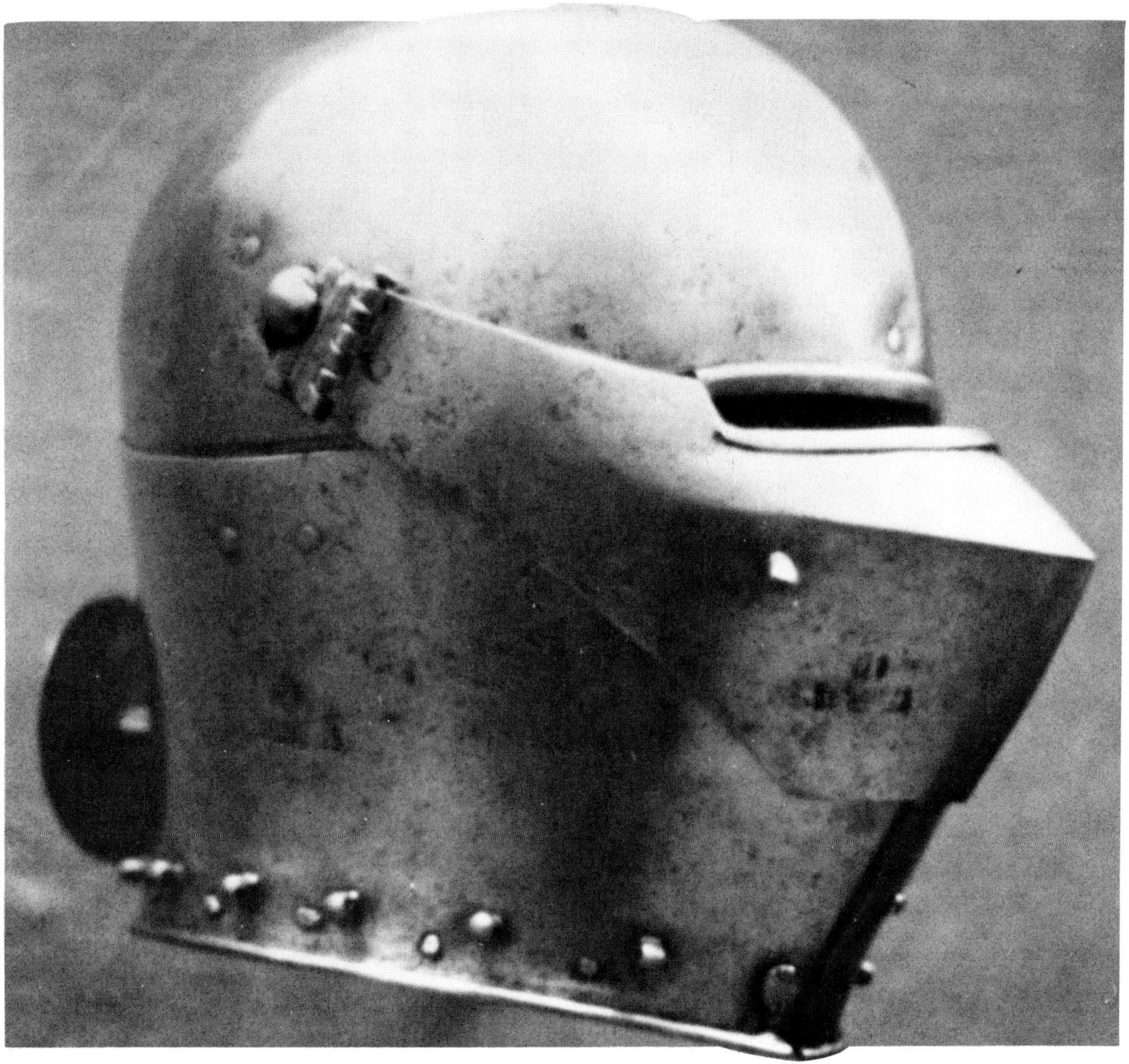

TALHOFFER'S FECHTBUCH,
Gothaer Codex, 1443, plate 57.

These foot combatants in the ring wear examples of the great bascinet with hinged visors and hooped armor skirts called tonlets. They prepare to do battle with spears, two-handed sword, and dagger, each event with a limited number of exchanges. Note the short-sleeved hauberk worn under the cuirass, protecting arm pits and groin.

ARMET (ELMETTO), ITALIAN (MILANESE), c. 1450.
Tower of London Armouries (IV. 498).

This helmet is struck twice with an indistinguishable mark, and came from the marvelous armory of the Trapp family at Churburg Castle in the Tyrol, midway between Italy and Germany. The armet, evolved from the bascinet, originated in Italy in the mid-15th century and was a great improvement in helmets, being neater, lighter; its weight was shifted from the wearer's head to the shoulders through the use of the gorget. This and the previous armet represent two of the earliest examples. The right cheek pieces on some bascinets were hinged on one side, as our example c. 1430 from the Metropolitan Museum, on page 91. As far as we know, the armet was the first type of close helmet in which both cheek pieces were hinged. It was always made of five pieces (skull, brow reinforce—riveted to the bowl, two cheek pieces and visor); the skull had a ridge of keel form, with a keyhole perforation for a crest at the top, and continued to the neck as a base for the seating of the cheek pieces; the occularium was the opening between the upper edge of the visor and the lower edge of the brow reinforce; the cheek pieces always at the top, the left overlapping the right and secured by a peg at the chin; and the visor was always of one piece hinged at the sides, sometimes secured by a hinge pin. On some Spanish armets, the visor extended upwards over the brow, and hinges were concealed.

GREAT BASCINET (BASCINETTO), ITALIAN, c. 1450. Kunsthistorisches Museum, Vienna, A-2.

This helmet was made by Tomaso Missaglia of Milan, in whose family were some of the finest craftsmen of their kind in 15th century Europe. Milan was a great center for armor and weapons manufacture from the end of the 13th century. By c. 1429 the Milanese armorers had developed the cap-a-pie armor with all its parts, accomplishing a sculptured effect in smooth steel. This helmet, from a complete harness (see following page), belonged to Frederick I, Elector Palatine 1425-76, ruler from 1451, who probably ordered it when he began his reign. It is struck with the marks of Tomaso Negroni da Ello (Missaglia), his son Antonio, Master Antonio Seroni, and Pier Innocenzo da Faerno. Antonio was the last of the clan to use the name Missaglia; his successors preferred Negroni or Negroli, the true family name.

COMPLETE GOTHIC HARNESS, ITALIAN, c. 1450.
Kunsthistorisches Museum, Vienna, A-2.

(See text, page 104.)

KETTLE HAT, (CAPELLO D'ARME, CAPPELLO DI FERRO), ITALIAN, c. 1450. Tower of London Armouries.

This type of helmet has seen a longer period of use than any other. It was used in Roman times and was popular in the medieval from the late 12th century. Although still used in this general shape in the 16th century, it evolved into the morion and cabasset at this time, and into the pikeman's pot helmets. There is a strong similarity between some of the modern day war helmets and the medieval kettle hat.

SALLET, (SCHALLER, SALADE, CELATA), GERMAN, c. 1450.
Tower of London Armouries (IV. 429).

This helmet came from the arsenal at Rhodes. Note the resemblance to some German helmets of this century. This shows a second style of visor slit being cut through the front of the one piece helmet.

BARBUT (BARBUTA, BARBUTE), ITALIAN, c. 1450. Metropolitan Museum of Art, New York.

A fine helmet from Chalcis Castle, it features an unusual movable nasal guard hinged to a bracket riveted above the face opening. (Sometimes a nasal guard was fastened in a fixed position with a threaded bolt.) In its popular form the barbut was apparently in use as early as the first half of the 14th century. Made from a single billet of steel, it was most popular in Italy from c. 1430 to the end of the century.

SALLET (CELATA), ITALIAN,
(PROBABLY BRESCIAN), c. 1460.
The Metropolitan Museum of Art, New York (23.141), Dick Fund.

The steel headpiece within the ornamental lion's head retains its original blued and painted surfaces and cushioned lining and has an articulated neck guard, a characteristic only used in the late 15th century. The embossed and copper gilt lion's head may have been an inspiration from Roman art depicting Hercules wearing the lion's skin or from the Etruscan god of Hades crowned with a lion's head. On the triumphal arch at the Castel Nuovo in Naples, which was erected by King Alfonso of Aragon-Sicily in 1470, there is a relief group of the King surrounded by his knights. The knight next to the King (the Crown Prince?) wears a lion-headed helmet practically identical to this one. This helmet has been called Venetian, but Brescia was the source of most Venetian arms and armor.

SALLET (CELATA), ITALIAN, (PROBABLY BRESCIAN), c. 1460.

Top Row: Left and Right side views.
Bottom Row: Rear and Front views.

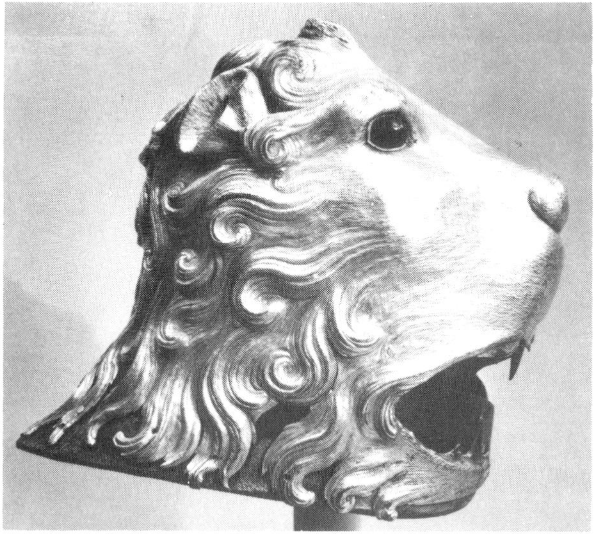

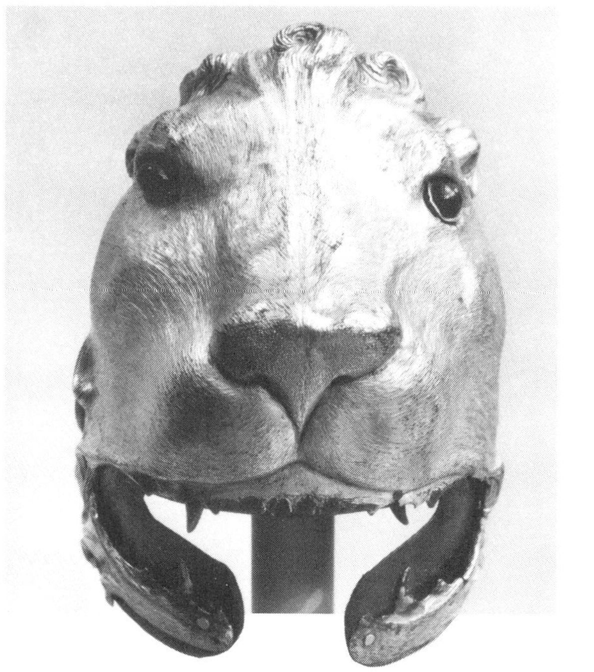

SALLET (CELATA), ITALIAN, (PROBABLY BRESCIAN), c. 1460.

The steel headpiece without its ornamental copper gilt lion's

BARBUT (BARBUTA), VENETIAN, c. 1450.
Glasgow Art Gallery and Museum, Kelvingrove, Glasgow.

The steel is covered with crimson velvet, retaining its original quilted lining of tow. It is believed to have belonged to Admiral Antonio Canal, commander of the Venetian fleet at the Battle of Negropont, 1450. He won the victory over the Turks, but was disgraced by the government because he kept back one galley which had his only son on board.

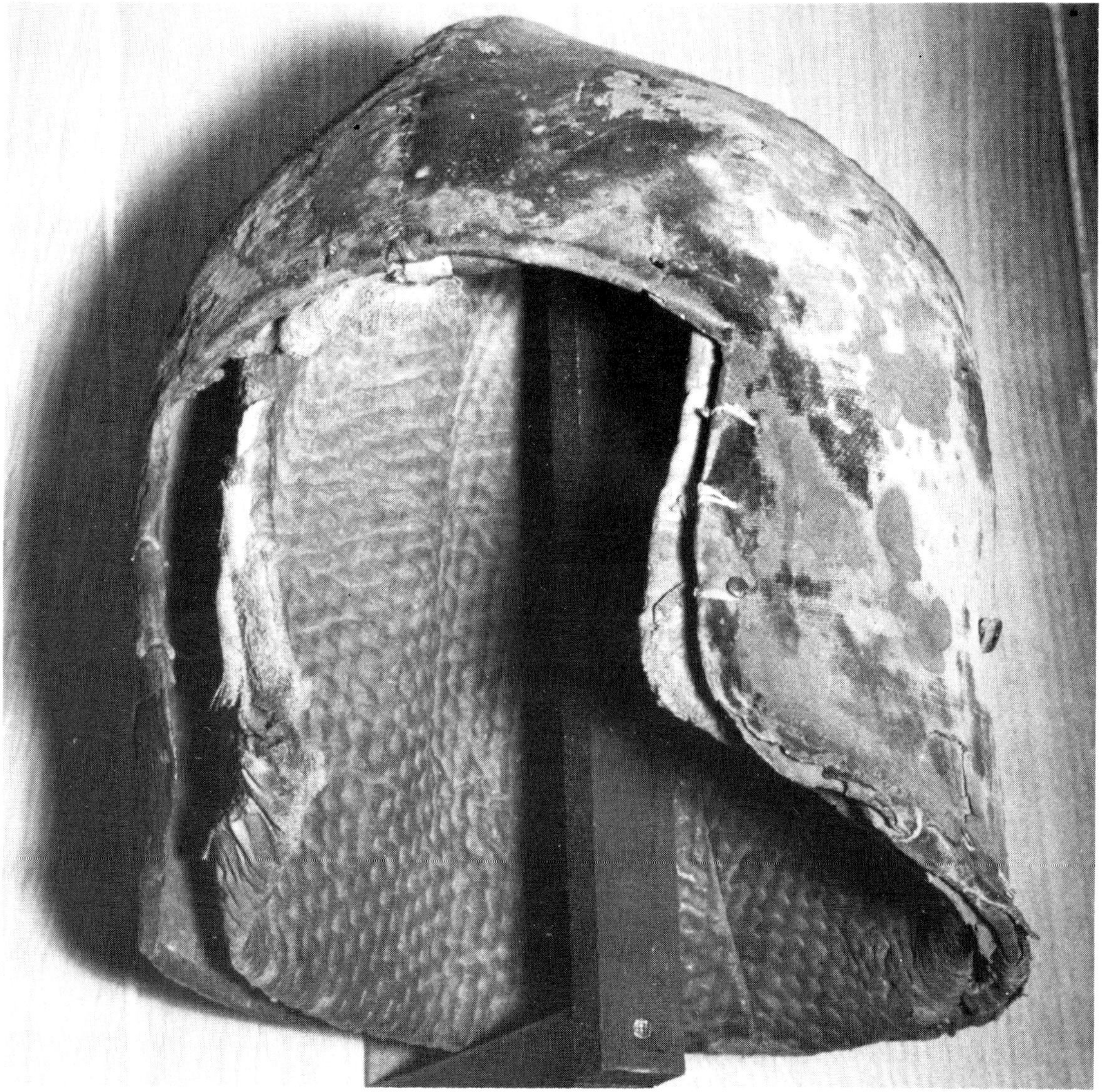

BARBUT (BARBUTA), ITALIAN,
SECOND HALF OF 15th CENTURY,
Glasgow Art Gallery and Museum, Kelvingrove, Glasgow.

Marked by the Missaglia family of Milan three times on one side, it also carries the mark of the Austrian Imperial Eagle, used by a patrician family in Venice. The barbuta allowed more vision and ventilation due to its open faced construction, and due to the Renaissance interest (rinascimento) in Graeco-Roman antiquity, was styled, in the 15th century, like helmets worn in ancient Greece.

BARBUTE, (BARBUTA), ITALIAN, 2nd HALF OF THE 15th CENTURY, Castel Sant' Angelo, Rome.

Struck on the back with Missaglia marks as well as that of the arsenal of Constantinople (a circle containing three vertical slashes, with what appears to be a composite bow resting horizontally on the longer central one). This was undoubtedly booty taken by the Turks after a battle. Note the lacing method of securing the lining. This indicates it was taken as booty after one of several battles the Turks fought with Italians in the 15th century.

KETTLE HAT, (CAPPELLO D'ARME, CAPPELLO DI FERRO), NORTH ITALIAN, c. 1460. Author's Collection, (245).

This well restored excavated piece was originally of one piece construction, with a narrow comb. It measures 9½ inches in height. Ex-collection of Sir Edward Barry of Ockwell's Manor, Bray, Berkshire, the president of the Meyrick Society preceding Sir James G. Mann.

BARBUT (BARBUTA), ITALIAN, c. 1460.
Tower of London Armouries (W. 17).

A goat's head and the letters ''IDB'' are punched twice under a split cross on the back (probably the de Bovis family of Brescia). Barbuts had either an open arch for a face opening (see previous barbut) or a ''T''-shaped opening such as this one (which resembled the classical Greek Corinthian helmets when the horizontal visor slit was oval shaped on each side). The ''T''-shape was discontinued c. 1470.

KETTLE HAT, (WAR HAT, CAPACETE), SPANISH OR FLEMISH, c. 1470, Tower of London Armouries.

This one has been struck with a crowned ''M'' mark. Notice the narrower brim as we come nearer to the 16th century. The keel shaped comb extending across the skull to the brim was characteristic of the second half of the 15th century.

TILT HELM (HEAUME CRAPAUD, STECHHELM) WITH CUIRASSE, FRENCH, c. 1470.
Mussée de l'Armée, Paris.

Forged in three pieces,
this is a fine French example
for the Gestech course (joust).

BARBUTA, MILANESE, c. 1470.
Formerly Marzoll Collection, Palazzo Marzoli, Brescia.
Author's Collection.

Central Keel-shaped ridge cut with a rectangular slot for mounting crest. Pierced at the border and around base of skull with circular holes for lining rivets. 10 inches high. The back of the skull struck on each side with the armorer's mark, a ''PR'' beneath a cross.

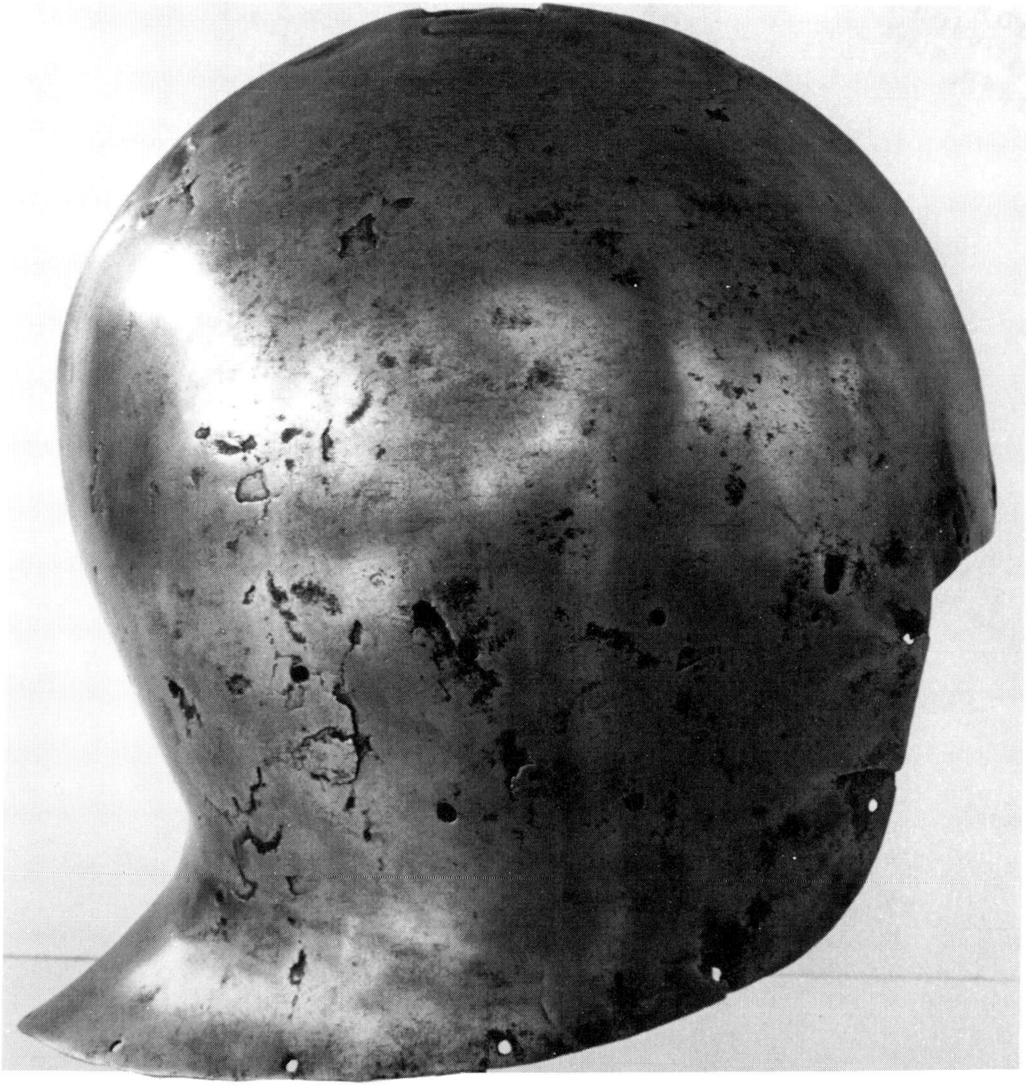

SALLET, (CELATA), MILANESE, c. 1470-80.
Author's Collection, (258).

A very light and mobile type, this is one of the excavated ones brought back to the United States from the Island of Rhodes (indirectly), by Bashford Dean. It was undoubtedly used in the battle between Mohammed II and the Knights Hospitalers of St. John of Jerusalem. Although won by the Knights, they suffered severe losses. Heavily rusted through, it was restored at the Metropolitan Museum of Art by Harvey Murton. The brow reinforce and neck guard, which Stephen Grancsay was gracious enough to furnish me with, were restored by R.W. MacWillie. This type had no visor slit and was worn tipped at whatever angle was desired.

KETTLE HAT (WAR HAT, CHAPEL-DE-FER), FRENCH, 1475.
The Metropolitan Museum of Art, New York.

A rare and beautiful exhibition of steel craftsmanship, of one piece construction. This one shows a fine bevor and gorget plate. (Bevors were hardly ever worn by Italians with their Elmettos.) The gorget was probably designed to be worn over a brigandine, to which the lower tip was anchored by means of laces. This helmet weighs 6½ lbs.

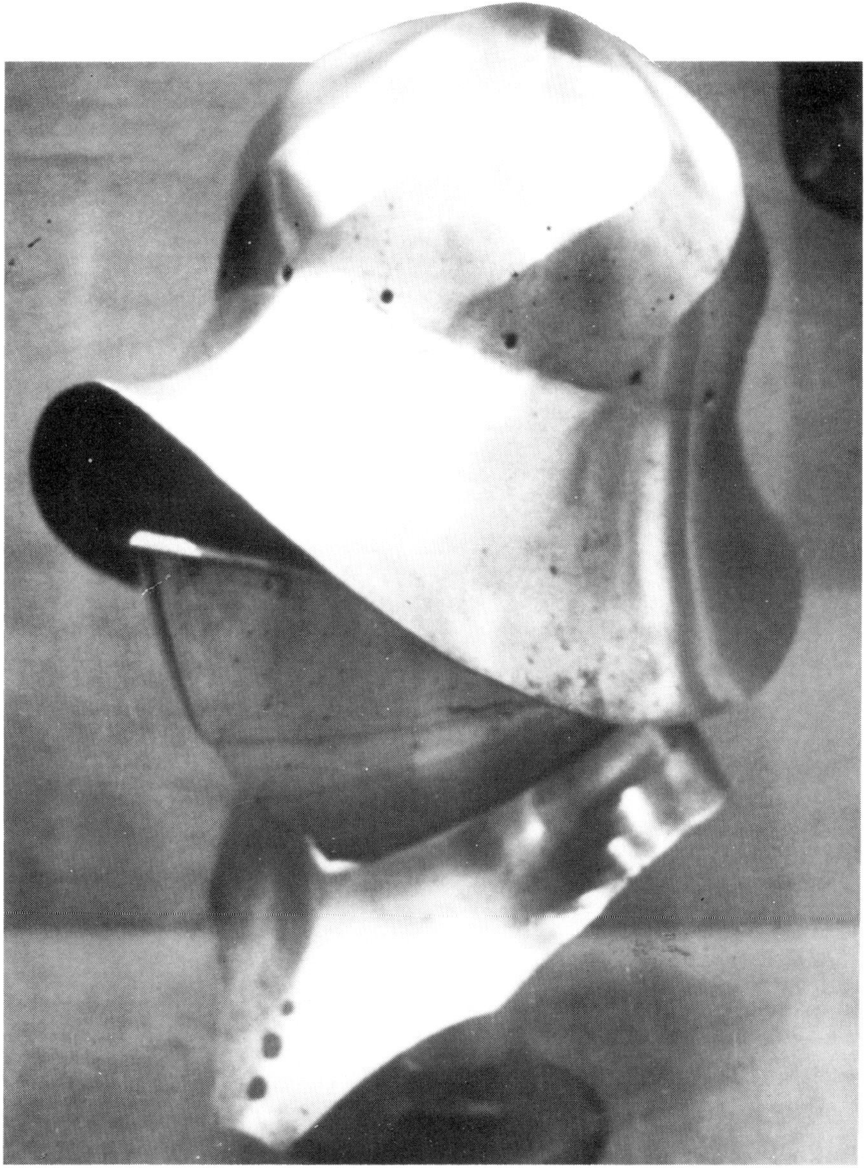

**SALLET (SCHALLER), PROBABLY GERMAN, c. 1475.
The Metropolitan Museum of Art, New York.**

This piece has a long pointed neck guard, giving it an appealing streamlined appearance. It is punched with a shield surmounted by a nine pointed crown, which is an arsenal mark, probably Burgundian, but possibly of Spanish origin, which would have been applied after it was exported to Spain; the movable visor has decorative cusped borders.

SALLET, ITALIAN,
THIRD QUARTER OF 15th CENTURY.
Metropolitan Museum Of Art.

Eight inches high with a strong medial ridge. It retains rivets for the lining around the base. From the Rhodes deposit of ancient armor abandoned by the Knights Hospitallers when the Island of Rhodes was captured by the Turks in 1523—a relic of the Christian wars. Excavated, it has restoration (little visible from this view) in 19th century style, riveted metal patches done under the direction of V. R. Bachereau, a Paris armor dealer, who purchased the Rhodes armor about 1880. His firm sold some of to Bashford Dean of the Metropolitan Museum of Art in 1920.

SALLET (SCHALLER), GERMAN, c. 1480.
The Metropolitan Museum of Art, New York. (14.25.580).
Riggs Benefaction, on loan to Los Angeles County Museum.

All one piece construction, the skull has a narrow flattened keel, flanged opening for vision, the lower edge protruding. The lower border of the helmet is turned outward, typical of most 15th century helmets. Weight, 6 lbs. 6 oz. There are ten rivets for securing the lining strap. Shown with a contemporary bevor.

SALLET (CELATA), ITALIAN, c. 1480.
The Metropolitan Museum of Art, 50.160 a.
Formerly Bashford Dean Collection.

Note the neck guard of two lames, brow reinforce, and the pivoted visor, vented and of "sparrow beak" shape, showing unusual step construction. This type of visor was more commonly used on armets. Around 1490 the bellows type visor was sometimes used on sallets. The vision slit is formed between the top of the visor and the bottom of the brow.

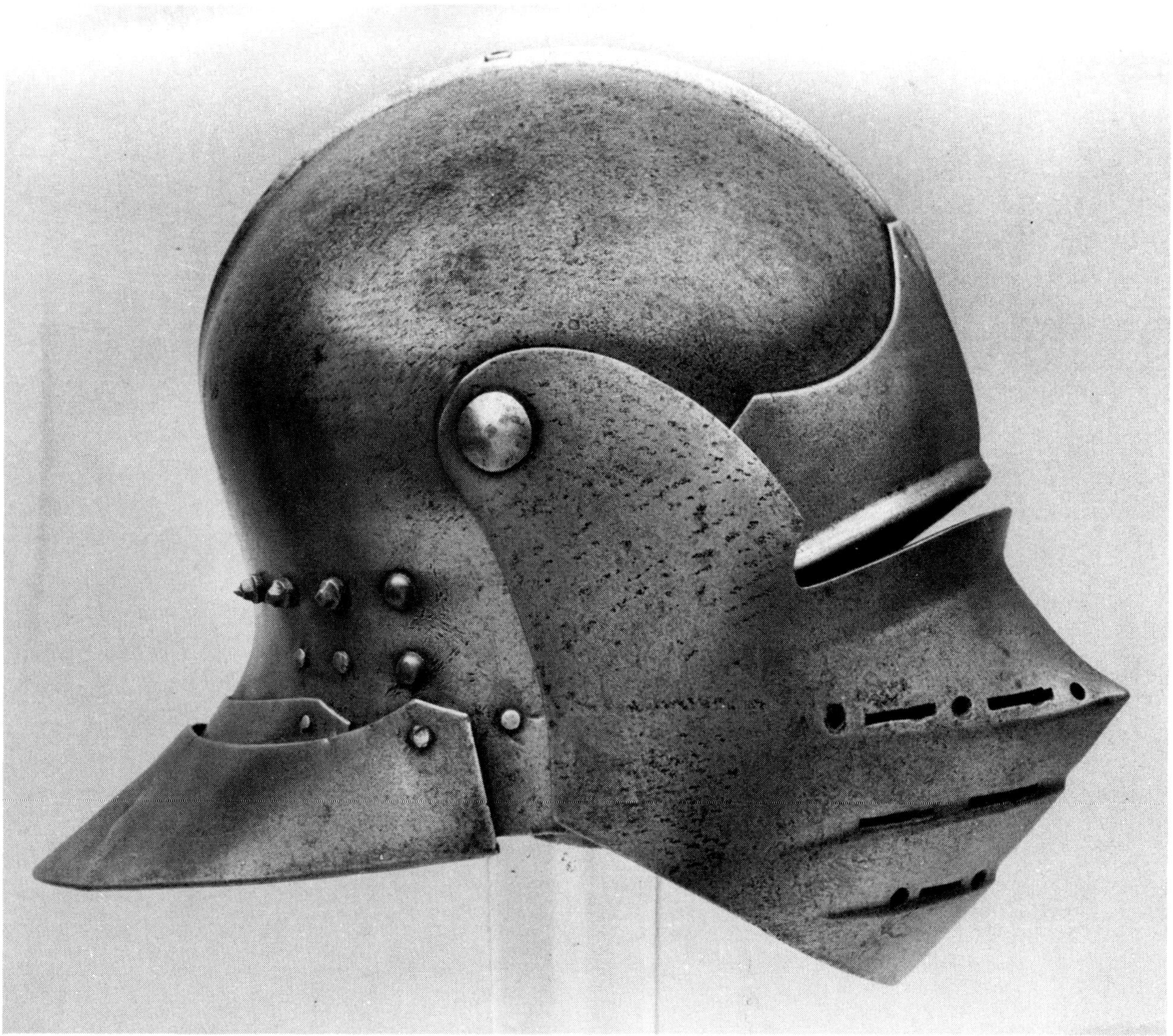

ARMET (ELMETTO ALL'ITALIANA, A BECCO DI PASSERO — SPARROW BEAK), ITALIAN, LAST QUARTER 15th CENTURY. Metropolitan Museum of Art, New York.

The rondel indicates use of a bevor. Note the reinforcing brow plate and the hole for a crest mount at the apex of the skull.

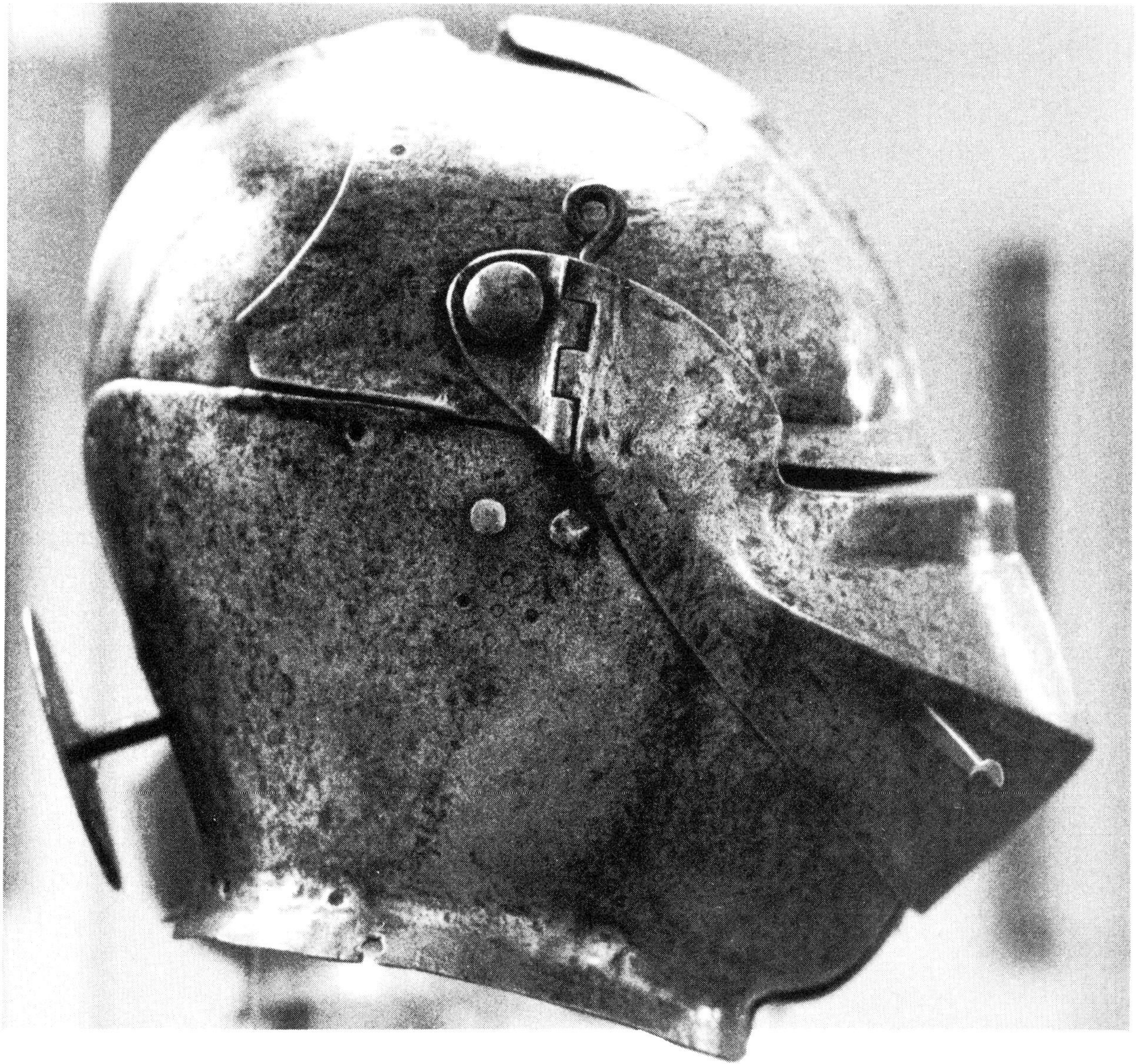

TILT HELM, ENGLISH, 1475-1500.
Glasgow Art Gallery and Museum, Kelvingrove, Glasgow.

This helm is composed of three pieces, the base plate added in the 16th century, perhaps for funeral use. The missing 3 x 4 inch ventilation door on the right was at one time closed with a spring catch. Weight 25 lbs. 13 oz. Tilt helms did not change style much during the second half of the 15th century, while helmets for war were subject to varying purposes and were also influenced by clothing fashions.

TILT HELM (STECHHELM), ENGLISH, c. 1480.
Tower of London Armouries (IV. 411).

This helmet weighs 22 lbs. 8 oz., and is known as the ''Brocas'' Helm, from a collection of that name. Helmets of this type, called ''frog-mouthed'' were apparently the first designed specifically for jousting. This is formed of three pieces of steel of different thickness, (thinnest in the back because of reduced exposure, thickest in front where most lance blows were taken) a characteristic of a genuine helm, and fixed by iron rivets ½ inch in diameter. Brass caps are soldered on the projecting rivet heads, a decorative technique frequently used on helmets in the 15th and 16th centuries. The front piece below the sight has been rolled back for added strength, eliminating the sharp edge, and a reinforcing plate is added to the upper plate just above the opening, on the inside. Eight holes are punched just behind the vertical line of rivets on each side, probably for ventilation, and added after the helm was initially completed. There are four pairs of holes in the crown for holding up the lining. The small piece missing from the lower left side was the result of a lance blow. Note the unusual hinged straps for buckling the helm to the breast and back plates.

KETTLE HAT (ELMETTO), ITALIAN, c. 1480.
Tower of London Armouries.

This came from the Arsenal at Rhodes and is probably Milanese.

MOUNTED COMBAT WITH REBATED SWORDS, SHOWING THE HEAUME DE TOURNO; (TOURNAMENT BASCINETS, GREAT BASCINETS), FRENCH, c. 1480.
A plate from Rene d'Anjou's "Le Livre des Tournois Du Roi René", Verve, 1946, Paris.

The figures represent combat between de la Gruthuyse and de Chistelle. After the required number of strokes with the sword, the baston (fluted wooden club seen at the right side of the horseman on the left) will be used. These tourneys, the Kolbenturnier, had a special characteristic in that no personal injury was intended the opponent, the object being to batter off the crest of his helm; purely a game of skill. The globose helm was latticed with strong iron bars and fastened to the cuirasse in front and back. It was heavily lined and covered with painted leather.

SALLET (CELATA ALLA VENEZIANA), VENETIAN, c. 1480-1500.
The Metropolitan Museum of Art, New York.
Bashford Dean Memorial Collection.
On loan to the Los Angeles County Museum.

This classic helmet, with its original red velvet covering and foliate metalwork extending from the border and across the crest, is of the open faced style and was worn by a high ranking officer.

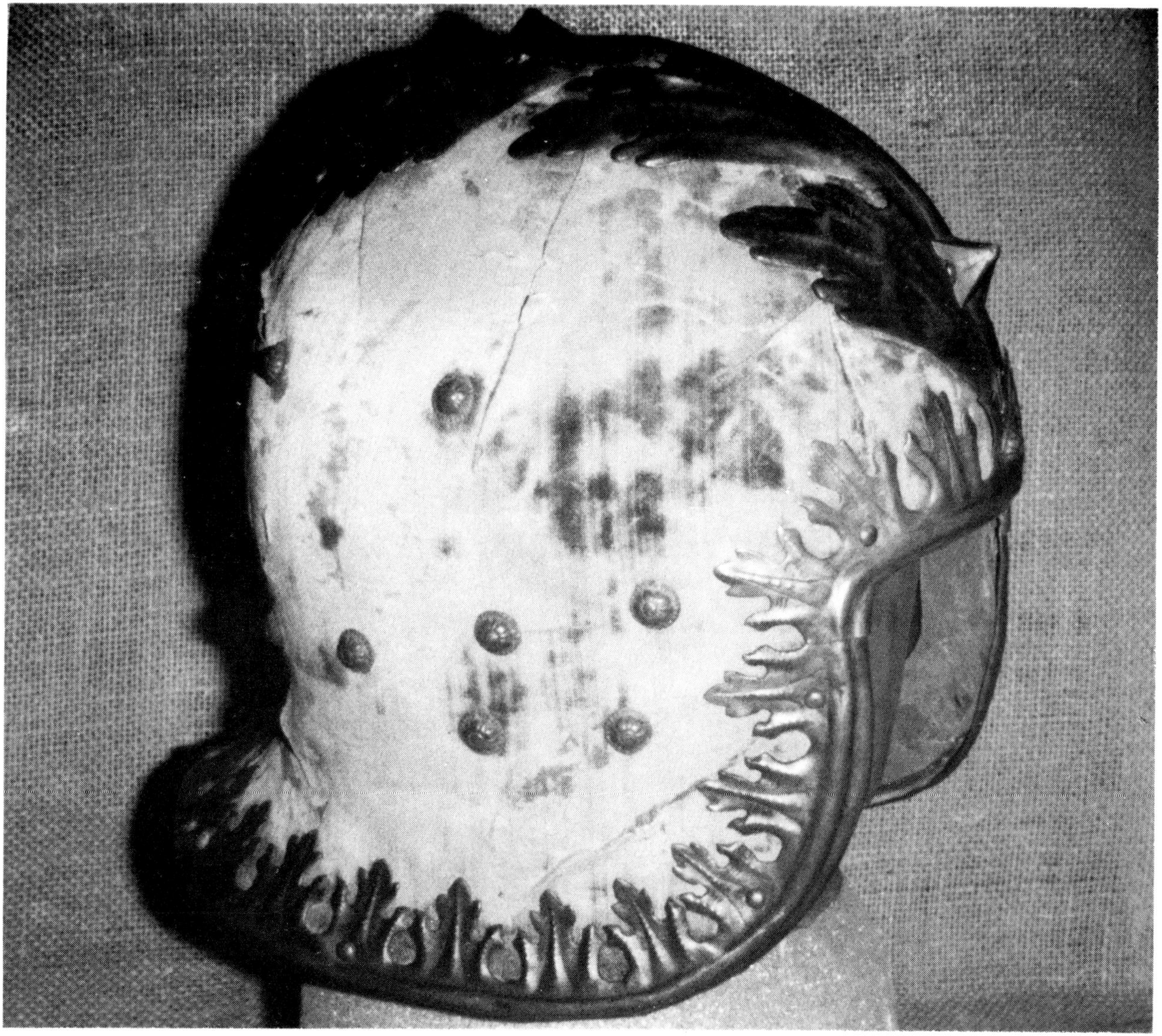

JOUSTING HELM WITH CUIRASSE, (CORAZZA DA GIOSTRA), ITALIAN, c. 1490. Palazzo Ducale, Venice.

For the Gestech. The visor is fixed by large bolts on either side below the skull piece. The back of the head-piece has fluted chevrons. The helm is attached to the breastplate by two lateral hinges and by a central strap with four holes to receive the staples attached to the breastplate. A securing bolt passes through the staples. The reinforcing breast-placate is attached by three large movable bolts, and on the shoulders with two adjustable straps. At the back of the skull piece are engraved three interlacing rings of the Sforza family (probably Cardinal Ascanio Maria Sforza, the Grand Elector of Pope Alexander VI — a man more prone to war and intrigue than church-like duties) and the initials ''A.A.'' It is also struck with a Milanese armorer's mark.

SALLET (SCHALLER), GERMAN, c. 1490.
The Metropolitan Museum of Art, New York.

Of beautiful simplicity in its one piece construction, its only decoration consists of fluted rivet heads for the lining strap. Helmets of this type were used primarily on horseback.

SALLET, (CELADA), SPANISH, c. 1490.
Photograph Courtesy Of Christie, Manson & Woods, Ltd., London. Author's Collection.

A field hat used by archers and the common soldier. This piece is seven inches high with a medial ridge. It is arched over each eye, with a vertical slit at ear level, a distinctive Spanish trait. The ear apparently was protected by a separated piece of leather, or iron, to afford better hearing.

ARMET-À-RONDELLE, (ELMETTO), ITALIAN, c. 1490.
The Metropolitan Museum of Art. 14.25.718.
Gift of William H. Riggs, 1913.

Of classic Italian form, it displays a separate collar from which is suspended the remains of a mail aventail.

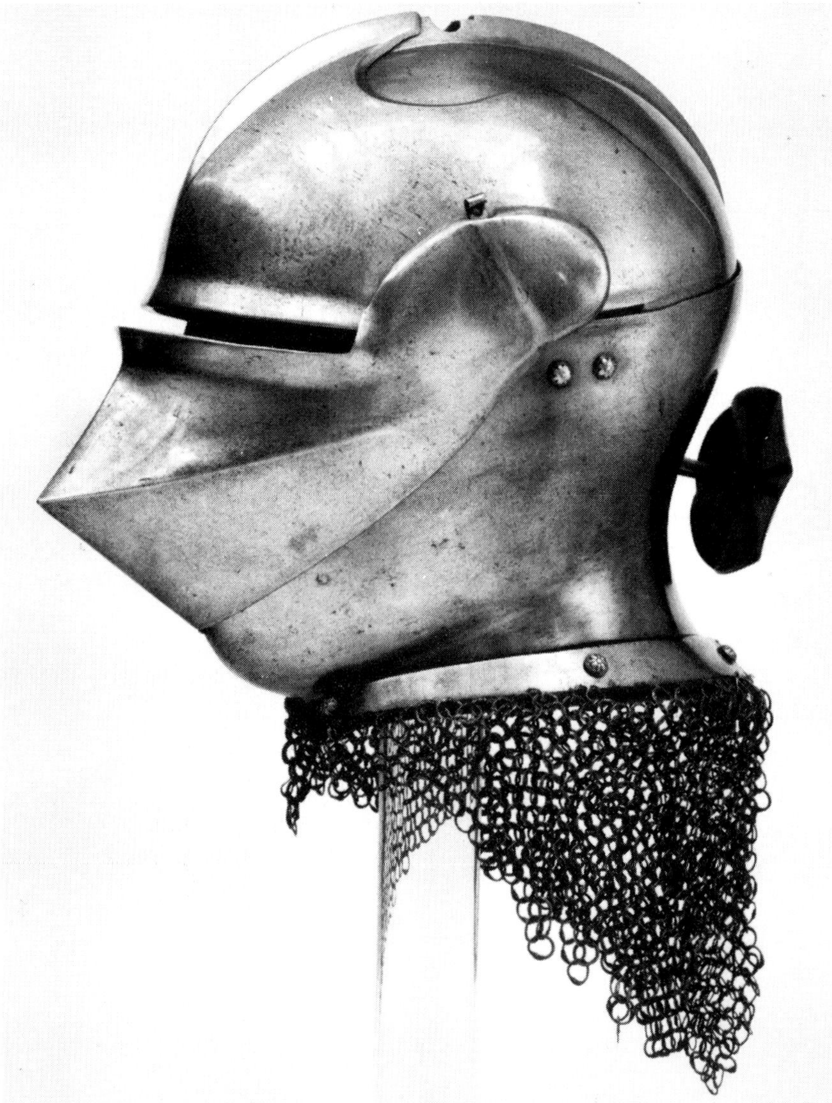

GREAT BASCINET (KOLBENTURNIER HELM), GERMAN, c. 1490.
Metropolitan Museum of Art, New York.
Gift of William A. Zabriskie, 1940. (40.135.3.).

Owned during its original period of use by the Von Stein family of Swabia, this fascinating head piece featured outstanding visibility and protection from the blunted club and rebated sword. A fancy horned crest, perhaps with a leopard sitting in the middle, probably accompanied this piece. Great Bascinets of a rounded shape utilized the high steel collar to replace the aventail. The type shown here was strictly for foot combat and tournament use. This one measures 18½ inches in height, and weighs 12 lbs. 10 oz.

GREAT BASCINET, (GERMANISCHES HELM), GERMAN, c. 1490. Kunsthistorisches Museum, Vienna (WSB23).

This type of helmet was designed for foot combat. The myriad of rectangular slits on the visor form an attractive pattern.

SALLET (SCHALLER), SOUTH GERMAN, c. 1490. Kunsthistorisches Museum, Vienna.

With pivoted visor and original painted decoration, this light style sallet was used by light horsemen or men at arms in general and was sometimes covered with fabric. There is seldom a turned edge at the bottom. They were known as ''black'' sallets because they were rarely polished bright. This helmet has been associated with the field harness of Roberto de San Severino, Count of Gajazzo, since the 16th century.

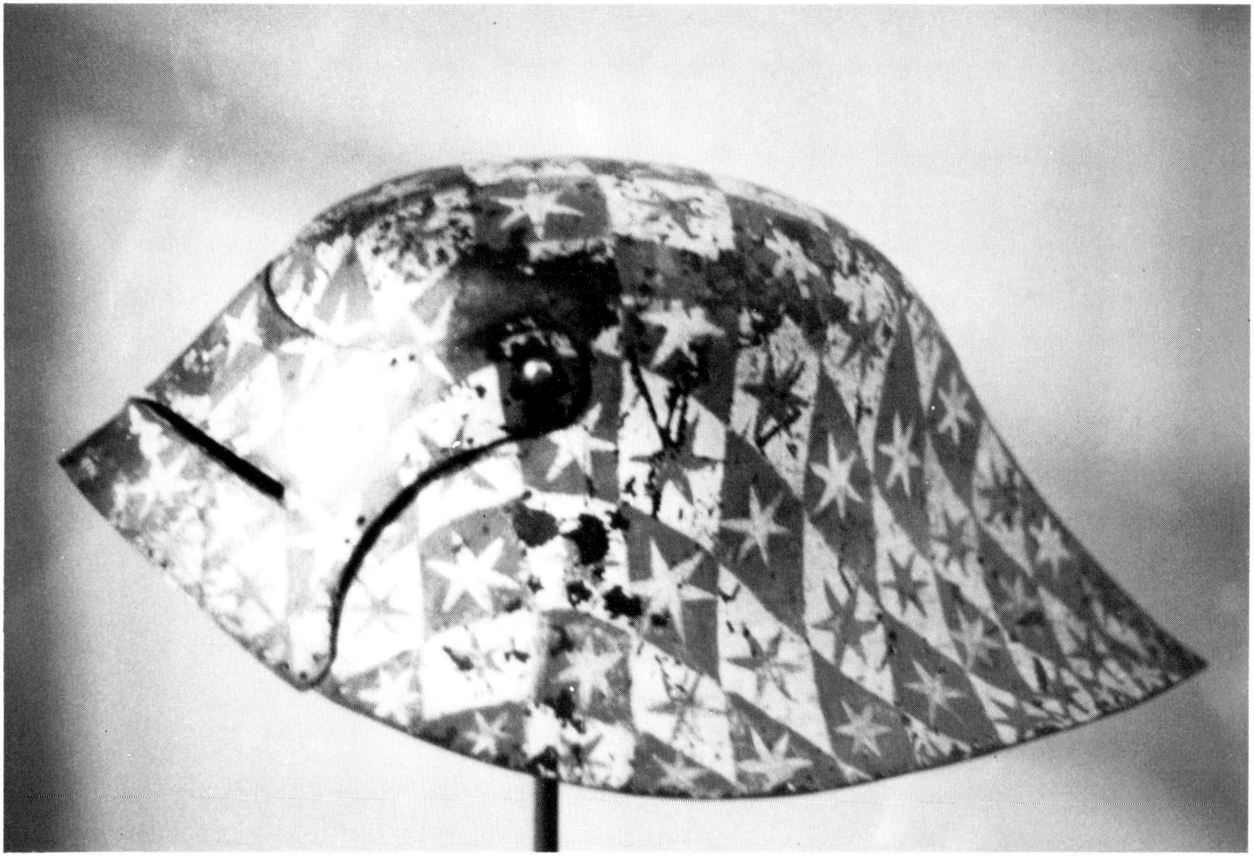

SALLET, PROBABLY ENGLISH, LAST QUARTER OF 15th CENTURY.
Private Collection.

This was excavated in Golden Lane, London, and was formerly in the Seymour Lucas and H. G. Keasby collections. Italian examples of these foot-soldiers sallets are quite similar in form.

WAR HAT, (CAPACETE), SPANISH, c. 1480-1500.
Tower of London Armouries (IV. 500).

Shown with bevor (Ill. 1300) protecting the chin and neck, mail shirt, brigandine (Ill. 47), and gauntlets (Ill. 1197-8); all Spanish except for the brigandine (probably German) of small rectangular steel plates riveted to a linen vestment. Note the skull shaped like half an almond shell; the small curved stalk at the apex and the pointed brim, front and back; typical of the Spanish cabacete, which was very popular. The bevor (barbote) worn with the cabacete had a separate hinged lame in the upper part which could be lowered by pushing a spring catch, visible in the photograph. Many of these cabacetes, as well as a good deal of armor and weapons, were manufactured by Italians in the style of the country to which they were intended for export.

SALLET (CELATA) WITH BEVOR,
ITALIAN, LATE 15th CENTURY,
Glasgow Art Gallery and Museum, Kelvingrove, Glasgow.

The pivot holes for the visor (missing) are evident on either side. Note the decoration on the rivet heads holding the brow reinforce plate. The associated bevor is possibly of German make.

KETTLE HAT WITH BEVOR OR BUFFE, (CHAPEL DE FER WITH BAVIÈRE), AND BRIGANDINE FOR AN ARCHER, FRENCH, LATE 15th CENTURY.
Musée de L'Armée, Paris.

An excellent example of armor for the common soldier, a great rarity because few of the plain examples have survived.

SALLET (SCHALLER), SOUTH GERMAN, c. 1490.
Royal Scottish Museum, Edinburgh.

For Scharfrennen (jousts with sharp lances), it has a reinforcing plate below the visor slit and decorative pewter washers for the lining lace holes. The holes pierced in the comb are for attachment of a crest.

TILT HELM (STECHHELM), GERMAN, c. 1500.
The Metropolitan Museum of Art, New York (29.156.67).

The helmet consists of three heavy plates: bowl, backpiece, and bevor (the latter deeply cut by lance thrusts), joined by steel, brass-headed rivets. It originally had a quilted lining cap, of which the only four original examples are in the Kunsthistorisches Museum in Vienna. The average thickness of the comb is 1/16 of an inch, backpiece 1/8 of an inch, and bevor 3/16 to 1/4 of an inch. There is a circular flange in the center of the comb for securing the crest. The engraved decoration on each side consists of hearts, diamonds, and floral design. Note the pierced hinge at the base of the backpiece for securing to the backplate. Formerly collection of Count Erbach-Erbach. Note also the restored lining straps. In the Gestech course the jouster attempted to unhorse his opponent and to splinter the lance, fitted with a coronel, against his opponent's breastplate. A normal saddle was used, but with a stechsack, a padded ''bumper'' hung around the horse's neck to protect his chest and the rider's legs.

ARMET, (VISIERHELM), GERMAN, c. 1500-05.
Castel Sant' Angelo, Rome.

This is not a close helmet but an armet, with its cheek pieces opening outwards at the front and hinged horizontally at the base of the skull on each side. The author was so impressed with the quality of this helmet that he has included both front and back views. The shaped hollow rim at the base of the helmet fastens over the top of the gorget or collar, allowing a rotary movement. This method was discontinued about 1550, with the exception of Greenwich armor, until the end of the century, when it was used on some jousting helmets. This armet is a rare example from a short period (1500-1505) when the armet began to be made again in Germany (probably at Innsbruck), in the Italian style in which cheekpieces were hinged at their upper edges. Although the earliest datable armet is on the German effigy of Dietrich Hofer (1416), Sünching, Bavaria, this type of helmet was not very popular in Germany from the first half of the 15th century, until the time the helmet was constructed.

ARMET, (VISIERHELM), GERMAN, c. 1500-05.
Castel Sant' Angelo, Rome.

Rear View.

SKULL CAP, (SCHALLER),
c. 1500-1520.
Author's Collection (212).

This simple helmet with its rudimentary tail is of finest quality; very compact steel, its hardness resulting from extra attention of the smith's hammer and is stamped with the mark appearing to be that of Wilhelm von Worms, the Elder, of Nuremberg. It probably was made for a Landsknecht officer who wore it with a half armor and a bishop's mantle of mail around his neck. Helmets of this type, because of their simplicity, are exceedingly rare.

ARMET (VISIERHELM), SOUTH GERMAN, c. 1500.
Tower of London Armouries (IV. 468).

The rondel at the back was probably intended to anchor the strap of a reinforcing wrapper (reinforcing bevor). This helmet carries the maker's mark of the Innsbruck craftsman, Hans Rabeiler, consisting of a sallet enclosed in a kidney-shaped shield set below his initials.

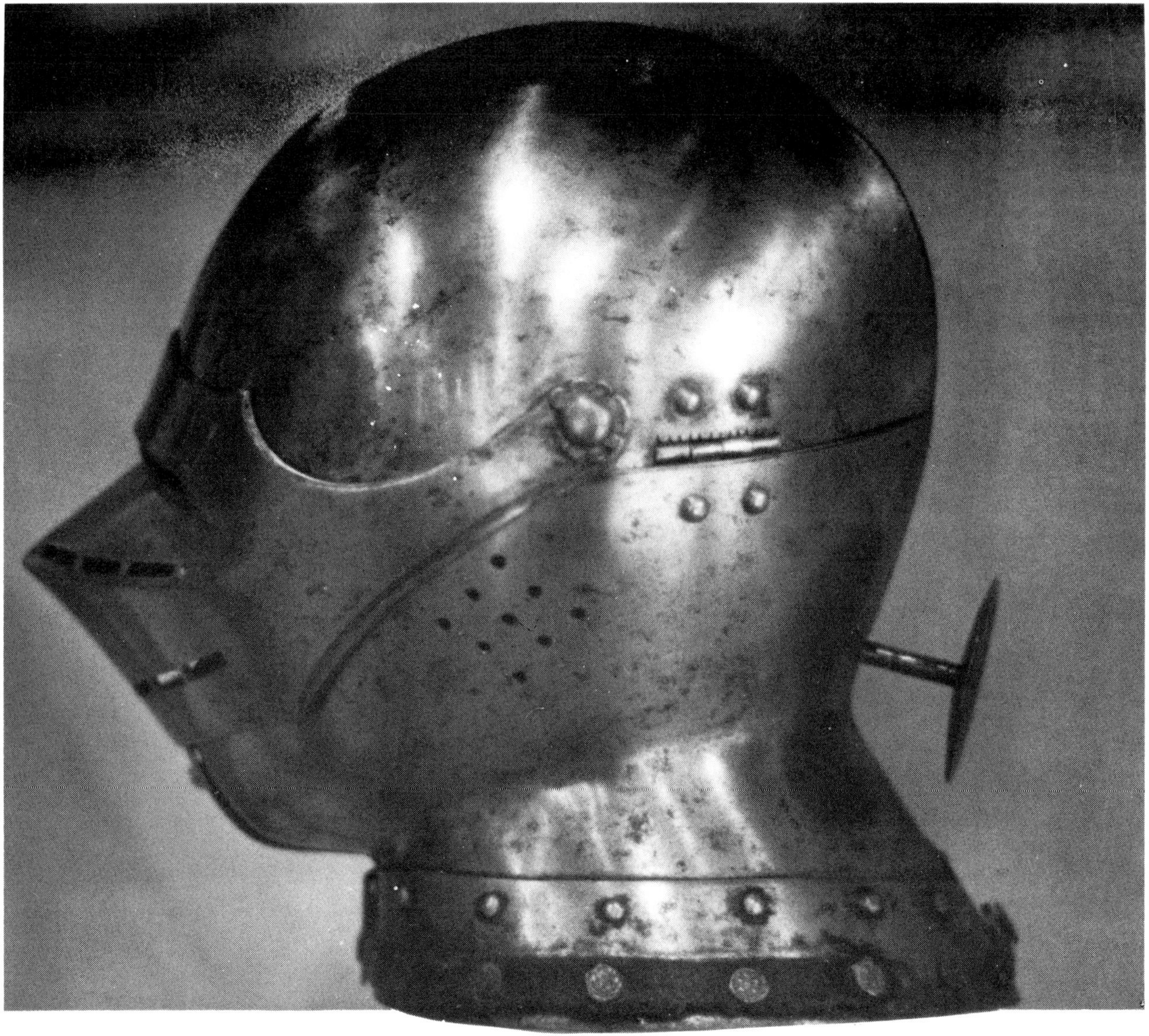

VISORED SALLET (GROSSE SCHALLER), GERMAN 1500-05.
The Metropolitan Museum of Art, New York. (29.158.35).

This fine piece is of the transition period from Gothic to Maximilian styles, the latter a merger of Italian influence and German Gothic fluting, possibly due in part to the inheritance of the Duchy of Milan by Emperor Maximilian. Note the three decorative raised ribs on either side, across the back of the skull. This head piece also represents one of the final stages in the evolution of the close helmet, having a pivoted bevor which the previous helmet of this type lacks. It was made by Hans Maystetter of Innsbruck; his mark consisted of an inverted "M" passing through a cross with an orb at the base. It has a wide median crest tapering toward front and back, with a perforation at the apex. A spring catch on the right holds the bevor closed; one on the left holds the visor closed. Visor and bevor are both pivoted at the sides from the same place. A hinged bar with a forked end is mounted on the right side. When fitted into a notch on the lower edge of the bevor, it held the visor and bevor open. The neck defence consists of three pivoted lames; the lowest has three sets of twin holes for lining laces and is punched with the armorer's mark.

ARMET (VISIERHELM), GERMAN, c. 1505.
The Metropolitan Museum of Art, New York,
Rogers Fund, 1904. (04.3.242).

While the previous example, page 187, has a visor of bellows form, this one displays the ''sparrow's beak'' shape. About 1505 another style of armet, with the hinges at the rear of the cheek pieces, was introduced and remained popular until c. 1560. The armet was used in England as late as the first quarter of the 17th century.

**BOY'S ARMET (VISIERHELM),
WITH BREASTPLATE AND FAULD, GERMAN c. 1500-10.
Tøjhusmuseet, Copenhagen.**

Period armor for a child is rare because it usually was worn out, and was not commonly produced. This one shows fine workmanship.

CLOSE-HELMET, (GESCHLOSSENER HELM), GERMAN, c. 1510.
Author's Collection. (259).

An early form of Maximilian styling, with fluted decoration, used c. 1500-1530. The earliest recorded example of a Maximilian harness is in Hans Burgkmair's woodcut of the Emperor, dated 1508. The one piece visor with twelve breath holes and six slots is of early bellows form, almost a sparrow beak style. Note the absence of a comb. The two rear necklames are replaced, and should not be fluted, but plain at this date. This displays true fluting which became evident about 1510. ''Maximilian'' armor is a convenient collector's usage and there is no proof the Emperor introduced the style. Armor with smooth finished surfaces was used equally with fluted armor in Germany during the Maximilian period, and was more popular elsewhere.

CLOSE-HELMET, (ELMETTO), FLEMISH OR ITALIAN, c. 1510.
The Metropolitan Museum of Art. 14.25.587.
Gift of William H. Riggs, 1913.

Note the cabled comb, combination sparrow beak and bellows visor, unusual neck guard of three lames (reminiscent of early German and Italian burgonets appearing in the next two decades); the rondel, usually applied to armets, and the strange guard dividing the visor slit. The strap from the bevor extending around the neck guard is an incorrect modern replacement. It should have been adapted to the rondel. The visor is not properly seated in the photograph, having been moved upwards a bit.

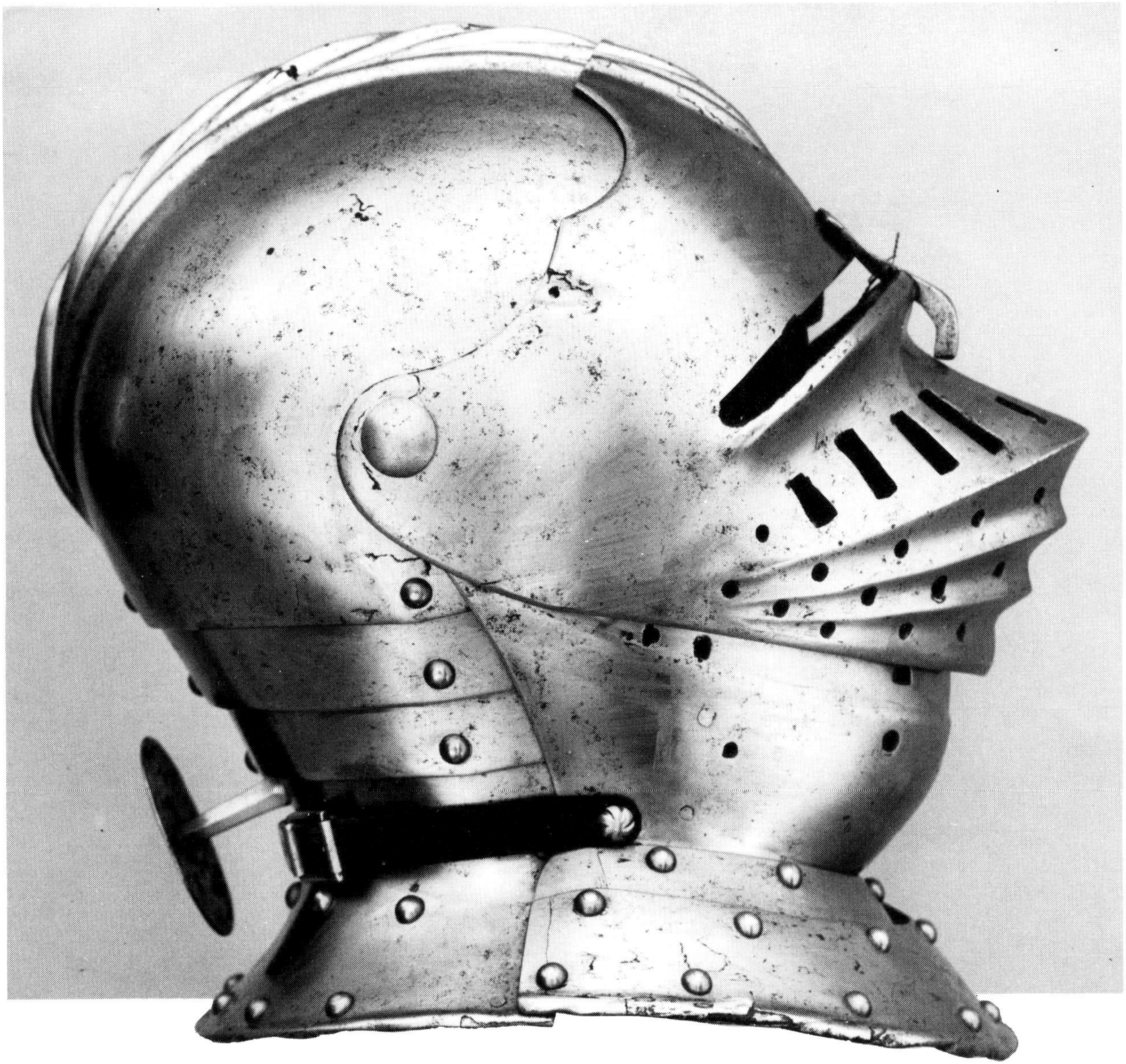

CLOSE-HELMET, (ELMETTO), ITALIAN, c. 1510.
Private Collection.

Note the lovely workmanship on this helmet, with its unusual fluting and attractive etching around the borders. Formerly in the Archduke Eugen Collection, this is a rare example of Italian work in the ''Maximilian'' style, illustrating the ''fusion'' of Italian and German styles which took place around 1500.

ARMET, GERMAN, 1511-14.
Tower of London Armouries (IV.22).

The sole surviving piece of armor from the one and only harness presented to King Henry VIII by Emperor Maximilian I, this grotesque helmet was made by Conrad Seusenhofer of Innsbruck. The mask visor is hinged at the top and anchors to the base of the cheek pieces, below the chin. The ram's horns fixed to the brow are of iron. The helmet has engraving over the eyebrows and cheeks, and was originally decorated with inlaid silver gilt bands. Weight 7 lbs. 8 oz. It should be noted that the horns are each skillfully forged from one piece of steel, as is the mask which possibly bears the likeness of Maximilian I.

CLOSE-HELMET, (GESCHLOSSENER HELM), SOUTH GERMAN, c. 1515.
Tower of London Armouries (IV-531).

Maximilian styling with fluted skull and bellows visor. Notice the spring catch for releasing the bellows visor, and another one to the left for releasing the bevor, both of which are pivoted at the same point. Instead of opening at the front of the chin (where the cheek pieces swing outward in order to don an armet), the close helmet breaks at the center of either side, the bevor and visor hinging forward, allowing it to be placed on the head.

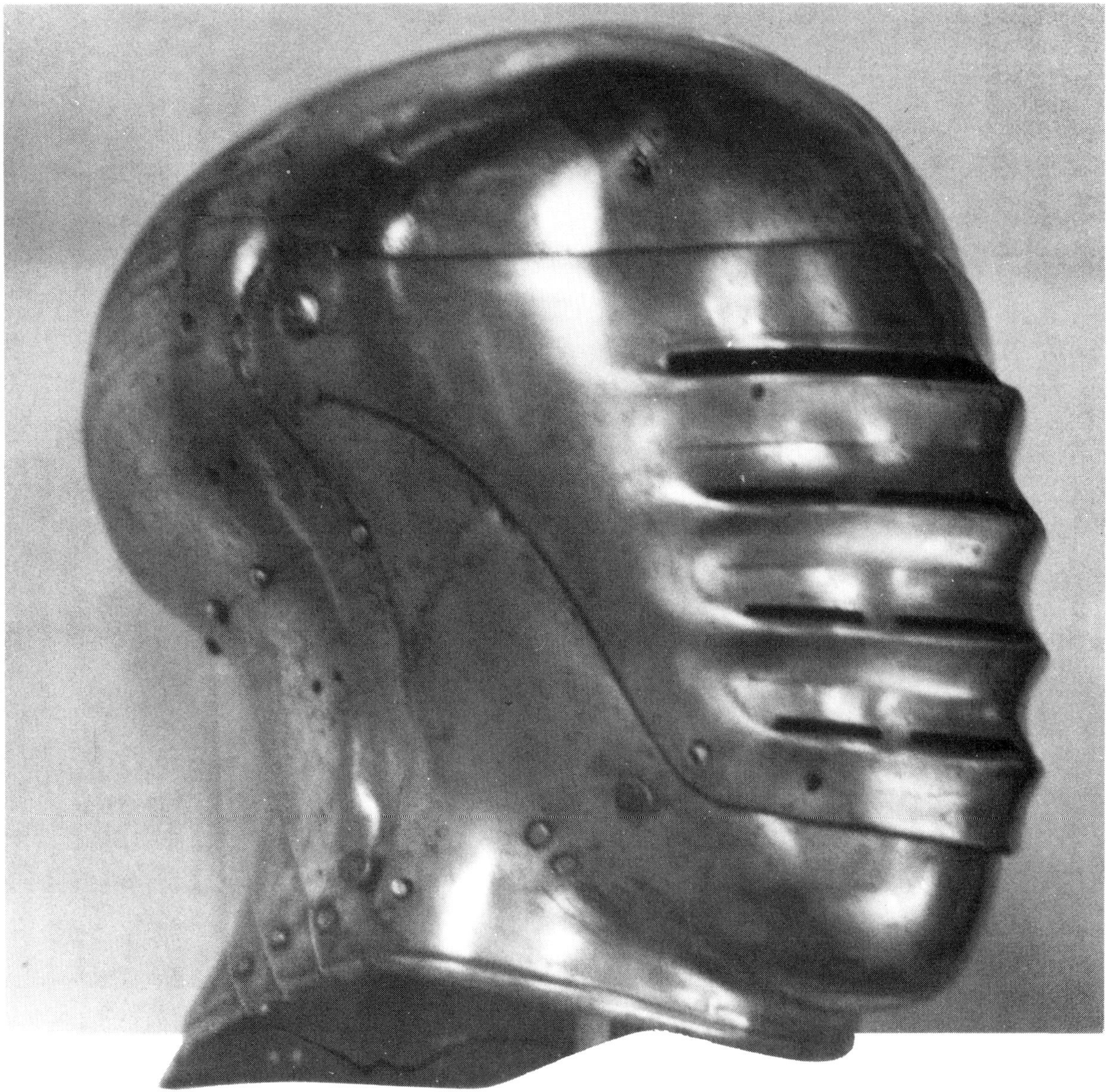

A GESTECH COURSE WITH CORONEL TIPPED LANCES, GERMAN (SAXON), c. 1527.
Eric Haenel, Der Sachsischen Kurfürsten Turnierbücher, Frankfort, 1910, Plate I.
(From the Turnierbuch des Herzogs Johann Friedrich des Grossmütigen, 1543, No. 81.)

Note the stechhelms and the lack of a barrier. The barrier, or tilt, was seldom used in Germany until the 16th century, but was prevalent on the rest of the continent from the first quarter of the 15th century, at which time jousting became a sport of its own rather than a training exercise for war.

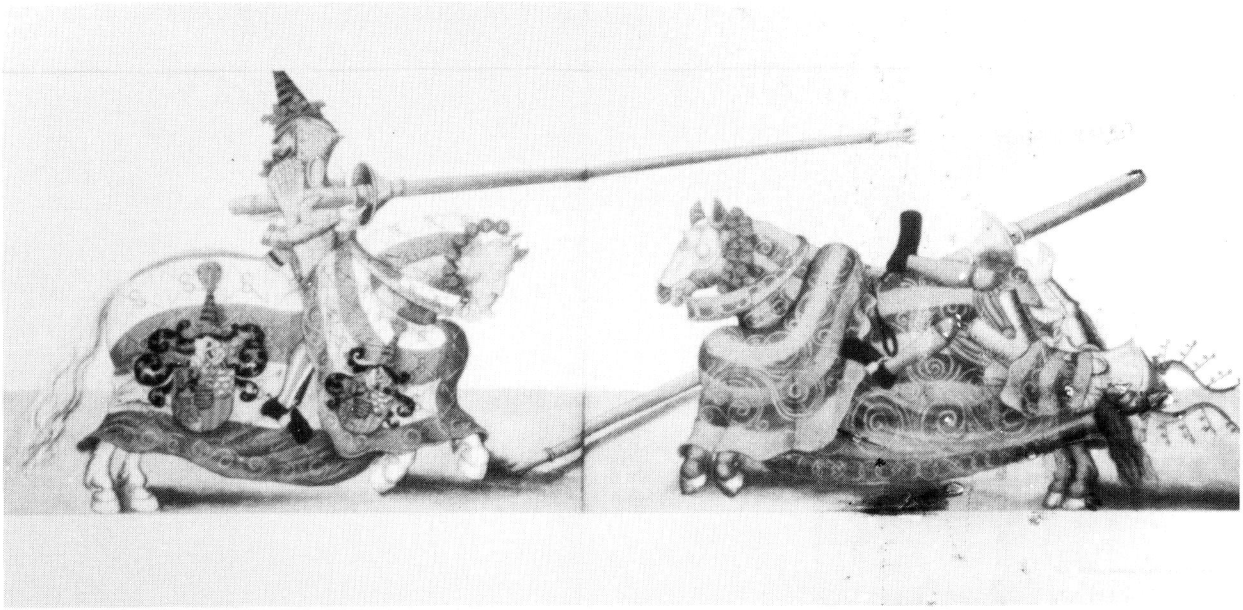

RENNHUT, (SALLET), GERMAN, c. 1520.
Tower of London Armouries (IV. 16).

Weighing 7 lbs. 14 oz. this helmet was designed for the Scharfrennen jousting course which used only a sharply pointed lance, the object being to unhorse one's opponent, while points were awarded for splintering of the lance. It was used with a bevor fastened by screws to the breastplate. The large button-headed rivets on either side behind the occularium were for attaching two wing-shaped re-enforcing plates which fell off when struck. The rennhut was made through mid-16th century.

CLOSE-HELMET, (GESCHLOSSENER HELM), GERMAN, c. 1520.
Tower of London Armouries (IV. 501).
Formerly W. R. Hearst Collection.

Probably an Augsburg helmet, this shows finely etched borders. Note the full development of the bellows visor. Formerly W.R. Hearst Collection.

CLOSE HELMET (GESCHLOSSENER HELM), GERMAN, c. 1520.
Tower of London Armouries (IV. 29).

This helmet with its grotesque visor in the form of a human mask with twisted moustache weighs 8 lbs. 4 oz. The skull is fluted with a cabled comb, and there is a cabled rim which locks over the gorget. Note the bolt and wing nut, an unusual method for locking the bevor in place. Grotesque helmets of this type were usually intended for parade use.

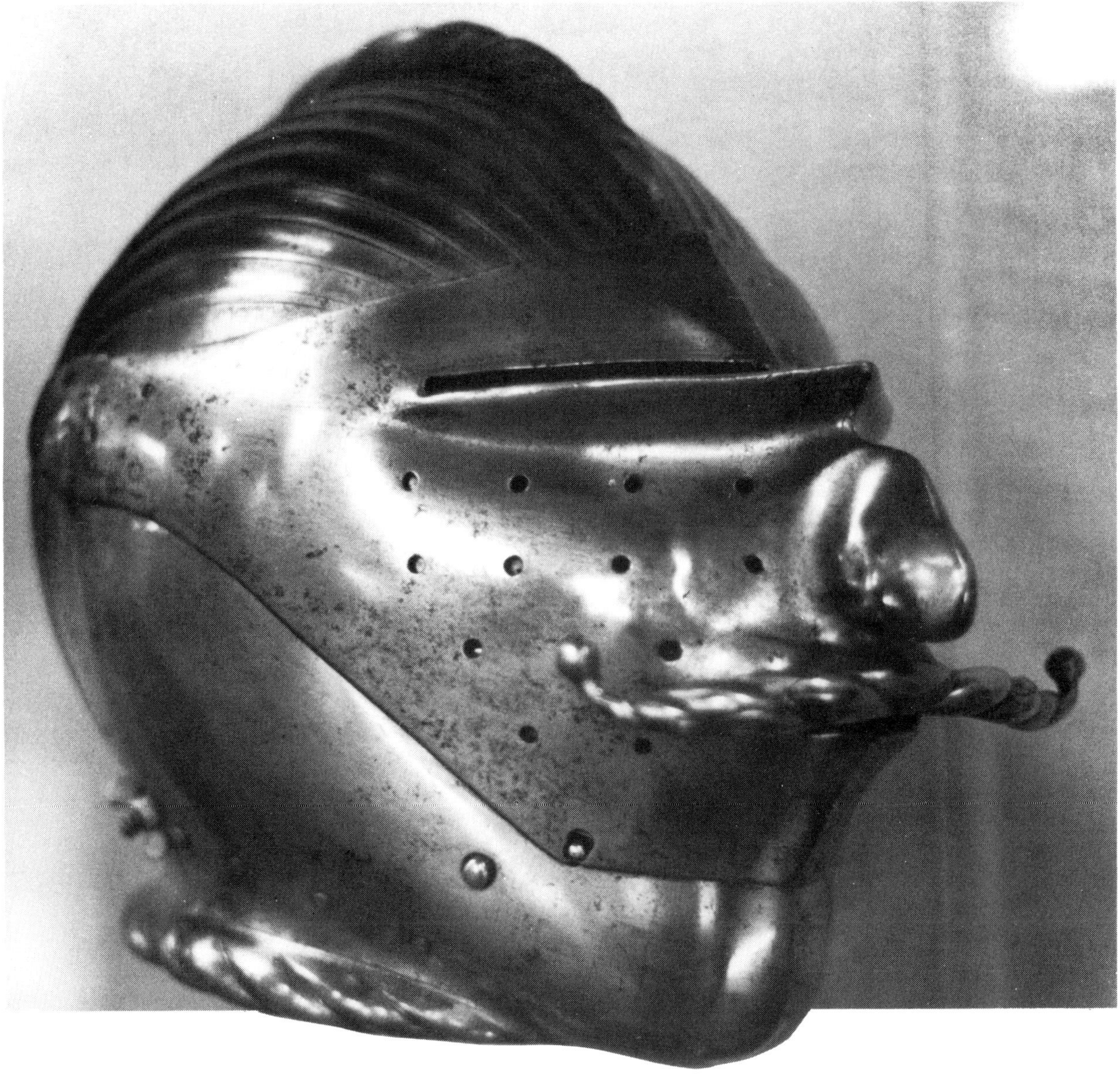

BURGONET (STURMHAUBE), GERMAN, c. 1520-30. Author's Collection (263). Formerly W. R. Hearst and Justin Stein, M.D. Collection.

This attractive helmet of unusual form is embellished with sunken margins, convex fluting and a heavily roped comb. The neck guard consists of three laminations for added flexibility. The rounded peak is hinged. The ear flaps are restorations by G.R. Donald, an armorer for W.R. Hearst. This represents an early example of the burgonet, its light weight making it desirable for both infantry and cavalry use, which began its appearance in Italy and Germany from about 1500. It was undoubtedly used by a Landsknecht armed with sword, pike or handgun.

KETTLE HAT, (EISENHUT, CAPELLO D'ARME), POSSIBLY GERMAN, OR OF ITALIAN MAKE FOR EXPORT TO SPAIN, EARLY 16th CENTURY.
The Metropolitan Museum of Art, New York
Rogers Fund, 1904. (04.3.236).

The 16th century Kettle hat was called a morion (prior to the appearance after 1550 of the popular, high combed morion with the brim curving up sharply to a pointed peak symmetrical in front and rear). This one has a low comb, down turned brim, and decorative tapered and fluted rivets for retaining the lining. Because it was open and didn't interfere with taking aim, it was a favorite with archers and musketeers. Kettle hats remained in use throughout the 16th century.

KETTLE HAT, (CAPELLO D'ARME),
PROBABLY ITALIAN, 1st HALF OF 16th century.
Norman Fargo Collection.

The helmet shown here retains the rounded skull and wide down-turned brim of the Kettle hat. It shows the transition to the classic morion with a well defined comb, of greater development than the narrow crest on kettle hats of the previous century. The open form of helmet was secured with broad leather chin straps, sometimes covered with riveted plates, and tied with laces under the chin.

CLOSE-HELMET. (ELMETTO), ITALIAN, PROBABLY EARLY 16th CENTURY.
Castel Sant' Angelo, Rome.

Note the barrel face guard (probably an adaptation of the 17th century) for the Pisan bridge tournament, Giuoco (del Ponte). Below are three targones; shields used also as the offensive weapon, and wielded in both hands by the two handles mortised onto the back side. They were decorated with colors of their owner's squadrons.

CLOSE-HELMET (GESCHLOSSENER HELM), GERMAN, c. 1530.
Private Collection. Photo courtesy of Ronald A. Lee and R.T. Gwynn.

An excellent example of craftsmanship from one of the most outstanding armour production centers of Germany, this helmet displays the Nuremberg mark just above the roped border on the right side.

CLOSE-HELMET, (GESCHLOSSENER HELM), GERMAN, c. 1530.
Private Collection. Photo courtesy of Ronald A. Lee and R.T. Gwynn.

Rear View:

BURGONET (STURMHAUBE),
SOUTH GERMAN, c. 1530.
Author's Collection (261).
Formerly J. R. R. Winsbury Collection.

The well formed one piece skull rises to an acorn finial; the hinged peak is struck with the Nuremberg mark, the half eagle to the left, chevrons to the right, within a shield.

HALF ARMOR, (ARMURE D'OFFICIER DE GENS DE PIED), FRENCH, c. 1530-60. Musée de L'Armée, Paris.

This belonged to Louis I of Bourbon, and is a pleasing example of etched and gilded decoration, the bourguignotte à visière (burgonet) has a pivoted visor and falling buffe, and is of exceptional quality.

CLOSED BURGONET (VISIERSTURMHAUBE), GERMAN, c. 1535.
Glasgow Art Gallery and Museum, Kelvingrove, Glasgow.
Formerly Bernal and Morgan-Granville Gavin Collection.

This fine helmet of Maximilian styling is possibly the Augsburg workmanship of Koloman Helmschmied. Designed for infantry use, the cheek pieces lock in the center below the barred face guard. This is fastened below by a pivot and catch and above by a series of hooks, so it can be raised with the pivoted peak.

PAGEANT HELMET, GERMAN, c. 1535.
The Metropolitan Museum of Art, New York.

This close helmet's visor slit is below the nose of the monster.

BURGONET, ENGLISH, c. 1535-40.
Private Collection.

Made in the Royal Workshop at Greenwich, during the reign of Henry VIII (1509-1547). Elements of armor made during the early years of the workshop, founded in 1515, are rare. Note the unusual one piece buffe, and the lower plate overlapping the skull at the back. Another unusual feature was a rod riveted to the buffe at top center that passed through a hole in the fall where it was secured by a hook (only broken parts remain). The incised parallel lines and the general shape are Greenwich traits.

HIGH-COMBED BURGONET, (STURMHAUBE), GERMAN, c. 1540-1550.
Author's Collection (226).

This helmet was believed to be French when it came from the Pierre Lorillard Ronalds collection. However, a rather similar one in the Tower of London (IV. 523) is suggested as being from Augsburg. The semi-circular peak with cabled edge blends in a straight line into the hinged ear flaps on either side, as does the neck guard (to the right). The high narrowly roped comb makes a lovely smooth transition into the sides of the skull. Of one piece construction, rough from the hammer and then painted black, this helmet displays a beautiful symmetry and shape. There are two holes through the neck guard for a porte panache (plume holder) now missing. Formerly Dr. Justin Stein Collection.

PARADE BURGONET (BORGOGNOTTA), ITALIAN, c. 1545.
Musée de l'Armée. Paris (H.253).

Forged from a single piece of steel by Filippo and Francesco Negroli. The crest is formed in the shape of a captive classical warrior whose beard is held on each side by his conquerors, Fame and Victory, each with the body of a winged griffin. In the center of the peak between the two griffins is a Greek inscription damascened in gold, which in translation means, "By them (Fame and Victory) I reach the stars." The lower border of the helmet is decorated with small damascened arabesques.

PARADE BURGONET (BORGOGNOTTA A BAVIERA), ITALIAN, 1545-50, Musée de l'Armée, Paris (H.149).

The work of Giovan Paolo Negoli for a member of the Medici family, it is embossed, partly gilded, and fashioned in the form of a dragon's head. The head piece has a removable upper bevor, embossed with a leaf mask that attaches in the dragon's mouth with turning pins. It is shown in place in the photograph. The Negroli family, famous for their embossed armors, were also noted for export armors and in the first half of the 16th century took over the reputation which this Missaglia family held previously in Milan.

VENETIAN PARADE BARBUT (BARBUTA ALLA VENEZIANA), c. 1550. Musée de l'Armée, Paris.

I have used the English term "barbut" (or "barbute") as opposed to the Italian "celata," because these helmets are more representative of the open-faced arched and "T"-shaped barbutas than the lower, more elongated "celatas." There is some confusion between the terms due to a type of sallet that "looks" rather half-way between the "vertical" barbuts and the "horizontal" sallets with pronounced tails. For practical purposes, if the helmet in question appears longer from front to back of the tail it could be called a sallet. This lovely helmet, covered with deep red velvet and its applied embossed metal work, crested with the Venetian lion, comes from the start of the period in which the classic Italian barbuta served as a parade rather than a battle piece. Barbuts of this type were made as late as the 18th century. It is possible some of these were used for funeral achievement. Also, some of these parade barbuts are actually headpieces from the late 15th century with decoration applied in the 17th or 18th century, used by the Doge's Guards at the Ducal Palace in Venice.

Casque de Parement
vénitien

BURGONET (STURMHAUBE), GERMAN, c. 1550.
Kunsthistorisches Museum, Vienna (WSA558).

Made for Kaiser Karl V, this beautifully embossed helmet represents some of the most outstanding artwork of Germany of the mid-16th century. It is embossed with scenes from the "Eneid," a late 12th century epic poem by Heinrich von Veldeke: there are Latin inscriptions on the sides; on the comb, Mars with his chariot; on the crown piece, the battle of Troy; and single battle scenes elsewhere. This is undoubtedly the work of Desiderius Colman of Augsburg.

HISTORIA EX LIB X VIRG
PVGNA RVTVLORVM
ANTIS

TRIPLE CRESTED BURGONET,
(DREIKAMIGE STURMHAUBE), GERMAN, c. 1550.
Tøjhusmuseet, Copenhagen.

An Augsburg mark (a pinecone) on the peak again points to the most outstanding center of armor craftsmen in Germany. It was closely followed by Landshut, Innsbruck, and Nuremberg. From the middle of the 15th century on, South German craftsmen, ''led'' by Helmschmieds of Augsburg, compared in quality with the Missaglias of Milan. Displayed with the burgonet are an unusually fine Landsknecht bishop's mantle (shoulder cape) of mail and a pair of elbow gauntlets.

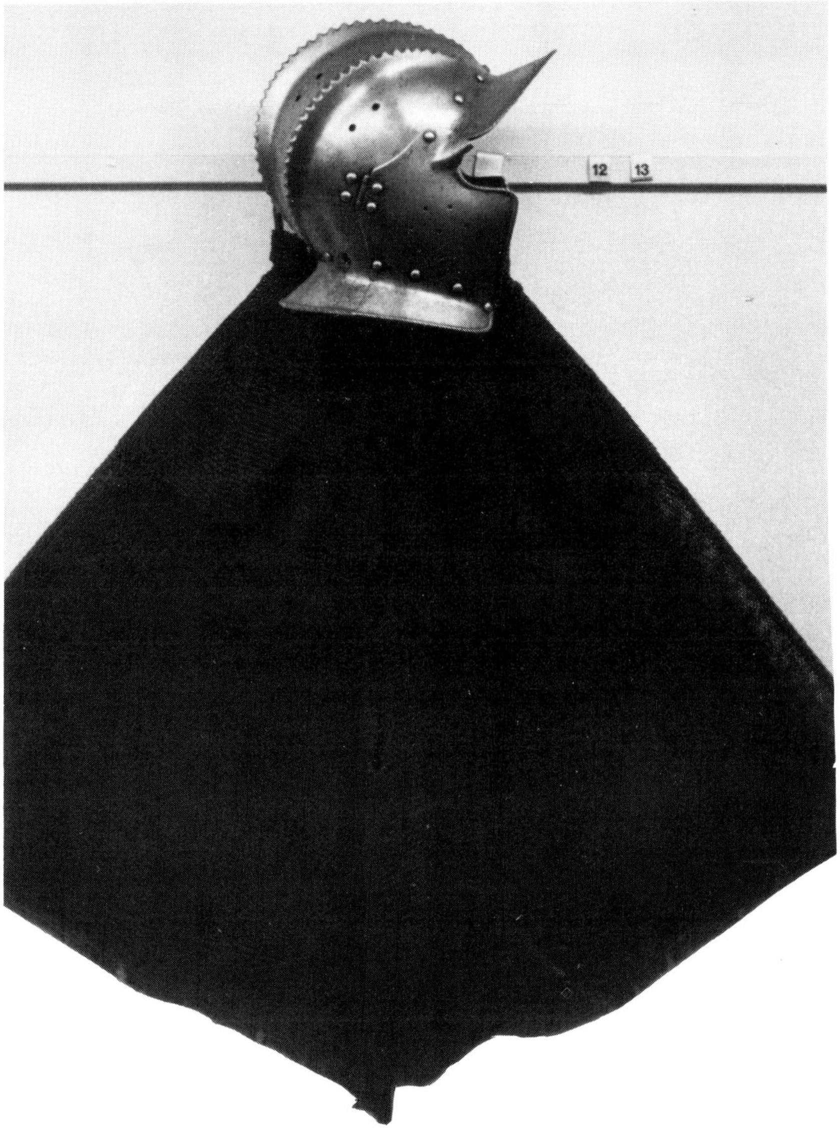
12
13

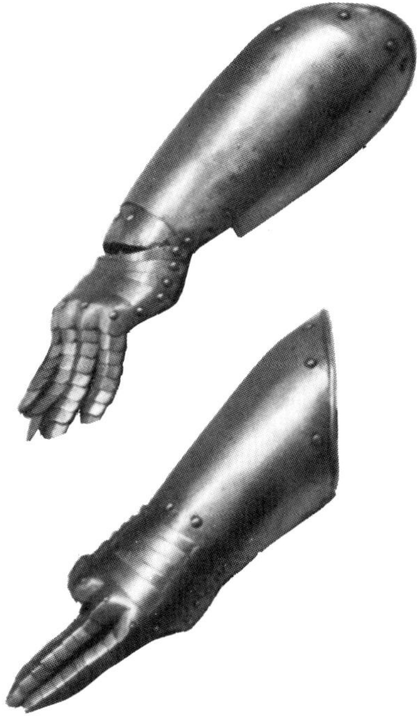

CLOSE-HELMET, (ARMET), FRENCH OR POSSIBLY FLEMISH, c. 1550. Musée de l'Armée, Paris.

This beautifully embossed work of art belonged to King Henry II of France. It shows the visor now divided into a visor proper and an upper-bevor below, and both are pivoted at the same place as the main bevor. The term "armet" has been retained for a close helmet to the present day in France. The full French term, **armet a gorgerin,** does not technically apply here, as the gorget is missing.

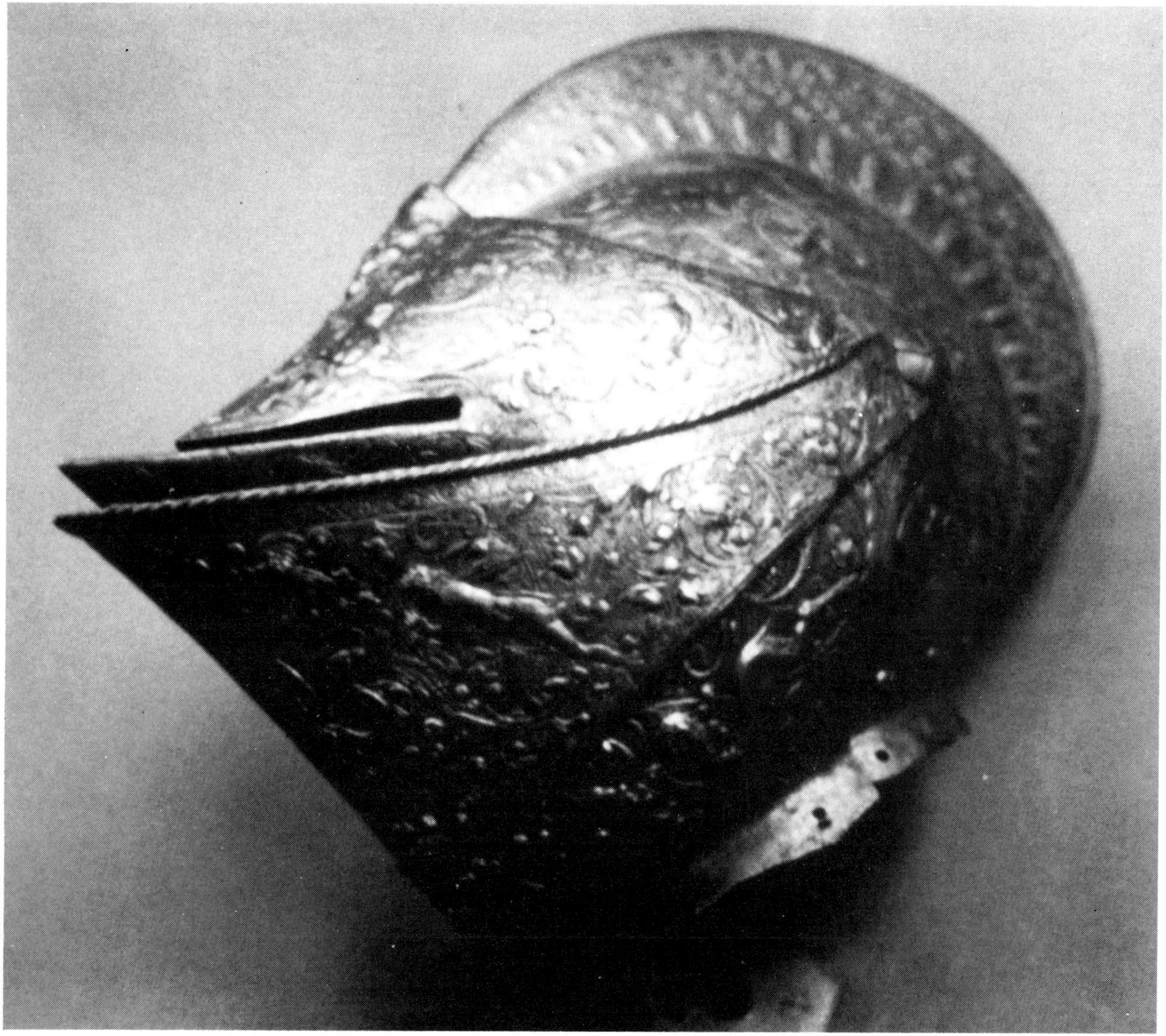

BURGONET (STURMHAUBE),
GERMAN, MID-16th CENTURY.
Glasgow Art Gallery and Museum, Kelvingrove, Glasgow.

Embossed with floral scrolls, this was for use with a black and white armor of quality. Its workmanship suggests Innsbruck manufacture.

BOYS CLOSE-HELMET, (ARMET A GORGERIN), FRENCH, MID-16th CENTURY. Musée de l'Armée, Paris (G. 188).

Russeted and gilded, this lovely armor is decorated overall with quatre-foils. (Its leather straps are covered with red velvet.) The close helmet, lined with purple satin and bordered with gold, carries a porte-plumail (plume holder) at the right rear.

BURGONET (STURMHAUBE),
GERMAN, MID-16th CENTURY.
Glasgow Art Gallery and Museum, Kelvingrove, Glasgow.

I have chosen this, along with many of the helmets in this volume, because of its esthetic appeal. Its four-sided skull tapering to a pyramidal finial, cabled borders and pivoted peak with flower petal rivet caps, blend to make this relatively simple head piece pleasing to the eye. Note the unusual plume holder to the rear of the skull.

PARADE BURGONET (BORGOGNOTTA), ITALIAN, MID-16th CENTURY.
Musée de l'Armée, Paris (I. 257).

A large oval medallion embossed in front center of the skull represents Pomone, the ancient Roman goddess of fruit trees. The peak displays two birds sampling fruit, a warrior on each side in classical harness. Embossed on the sides are figures of Mercury, Saturn, Diana, and Neptune. Jupiter and Juno sit back to back on the neck-guard, and on the crest is a mythological figure with a lion's head.

PARADE CASQUE, (BOURGUIGNOTTE), FRENCH, MID-16th CENTURY.
Musée de l'Armée, Paris (H. 254).

Damascened and gilded, the crest is surmounted with a dragon whose feet rest on a caryatid. Heavily embossed fauns, mythological figures and foliage make up the rest of the helmet. Its original lining is of green silk decorated with gold.

CLOSE-HELMET, PROBABLY ENGLISH, SECOND HALF OF THE 16th CENTURY, Author's Collection.

Of bright steel, this helmet has a one piece skull, high roped comb, roped lower edge to the visor, and two gorget plates to the front and rear. This helmet may have once been displayed above a tomb in an English church, as the comb is pierced for mounting a funerary spike. Note the brass caps to the lining rivets. Formerly Duke of Osuna Collection, Madrid.

PARADE BURGONET, (STURMHAUBE), GERMAN, 2nd HALF OF 16th CENTURY, Tower of London Armouries (IV. 154).

The surface is decorated with finely embossed designs of birds and fruit, with applied ornaments in gilt copper embossed with warriors and trophies.

CLOSE-HELMET, ENGLISH,
LAST HALF OF THE 16th CENTURY.
Victoria and Albert Museum, London.

This head piece is characterized by its roped and scalloped edges, stepped sights, prominent upper bevor, with its unusual forward reach, and its deep gorget plates. Laminated gorget plates were first added to the front and rear of the close helmet's base about the start of the second quarter of the 16th century.

BURGONET (STURMHAUBE), GERMAN, c. 1555.
Kunsthistorisches Museum, Vienna (WSA878).

This ornate example belonged to Archduke Ferdinand I and is styled in the Hungarian fashion. Ferdinand campaigned against the Turks as commander of the German army. The helmet is silver, decorated in oriental leaf ornament style. The peak, nasal, ear guards, neck guard, and lower part of the skull are etched. The lions' heads are the work of the Jamnitzer family of Nuremberg.

SALLET (RENNHUT), SAXON, 1550-1560.
Tøjhusmuseet, Copenhagen.

This Scharfrennen helmet belonged to Count Adolf, Duke of Gottorp, and was probably made by Hans Rosenberger of Dresden. Adolf was an heir to the throne of Norway and was presented with the Order of the Garter after his stay in England in 1560, where he proposed to Queen Elizabeth. This lovely one piece helmet with its fine etching, once gilt, is probably one of the last ''authentic'' rennhuts forged until the Gothic revival in the 19th century.

CLOSE-HELMET WITH GORGET, (ELMETTO A GUARDACOLLO), ITALIAN, c. 1560.
Author's Collection (246).
Formerly W. R. Hearst and Justin Stein, M.D. Collections.

Decorated with etched strapwork and classical figures, this piece retains traces of original gilding.

HIGH COMBED BURGONET (STURMHAUBE), GERMAN, c. 1560.
Author's Collection (262).

This highly tempered bright steel helmet with its etched brass plume holder represents simplicity, yet, because of its fine lines, meets the eye with a good deal of pleasure. Its fine workmanship was probably that of Anton Pfeffenhauser of Augsburg. It was in the Archduke Eugen Collection at Hohenwerfen fortress, near Salzburg, Austria, before it was acquired by W.R. Hearst. Archduke Eugen was a descendant of the Emperor Maximilian I, whose interest in quality armors and love of swordplay and the joust kept chivalry alive in the early 16th century.

A SCHARFRENNEN COURSE WITH POINTED LANCES, GERMAN, (SAXON), FEBRUARY, 1566, IN DRESDEN. THE FALLEN RIDER IS HANS CASPARN VON RÜXLEBEN. From the Turnierbuch des Kurfüsten, August, N. 54.

The riders wear a one piece Sallet or Rennhut, with a deep bevor fastened to the breast with screws. The large shield is of wood and leather reinforced with iron and is called a Renntartsche. Armors of this type were made at least until the 1550s, their use continuing a while after that.

Ein Rennen gethan mitt

CLOSE-HELMET WITH GORGET, (ARMET A GORGERIN), FRENCH, c. 1570. Musée de l'Armée, Paris (6.19).

The armor of Francis II, engraved and gilded in a fashion imitating brocade; probably from the Louvre workshop.

SKETCH OF A FIELD CLOSE-HELMET OF GREENWICH ROYAL WORKSHOP MANUFACTURE, c. 1570, Tower of London Armouries, (IV. 41).

a. gorget plates
b. spring catch for lower bevor
c. lower bevor
d. pivot hook
e. spring catch for upper bevor
f. upper bevor
g. ventilation holes
h. roped border
i. stepped sight
j. visor
k. comb
l. pivot
m. skull
n. plume holder (not present on this specific helmet)

The ventilation holes shaped as vertical slits with a hole in the center and the concave profile below the point of the visor are distinctively of Greenwich character. Henry VIII established the Almain Armoury in 1515 at Greenwich by importing eleven German and Dutch craftsmen (Almains), who were employed to make quality armor for himself and his court (or those with a royal warrant). Henry hired armorers from Italy and Brussels as early as 1511.

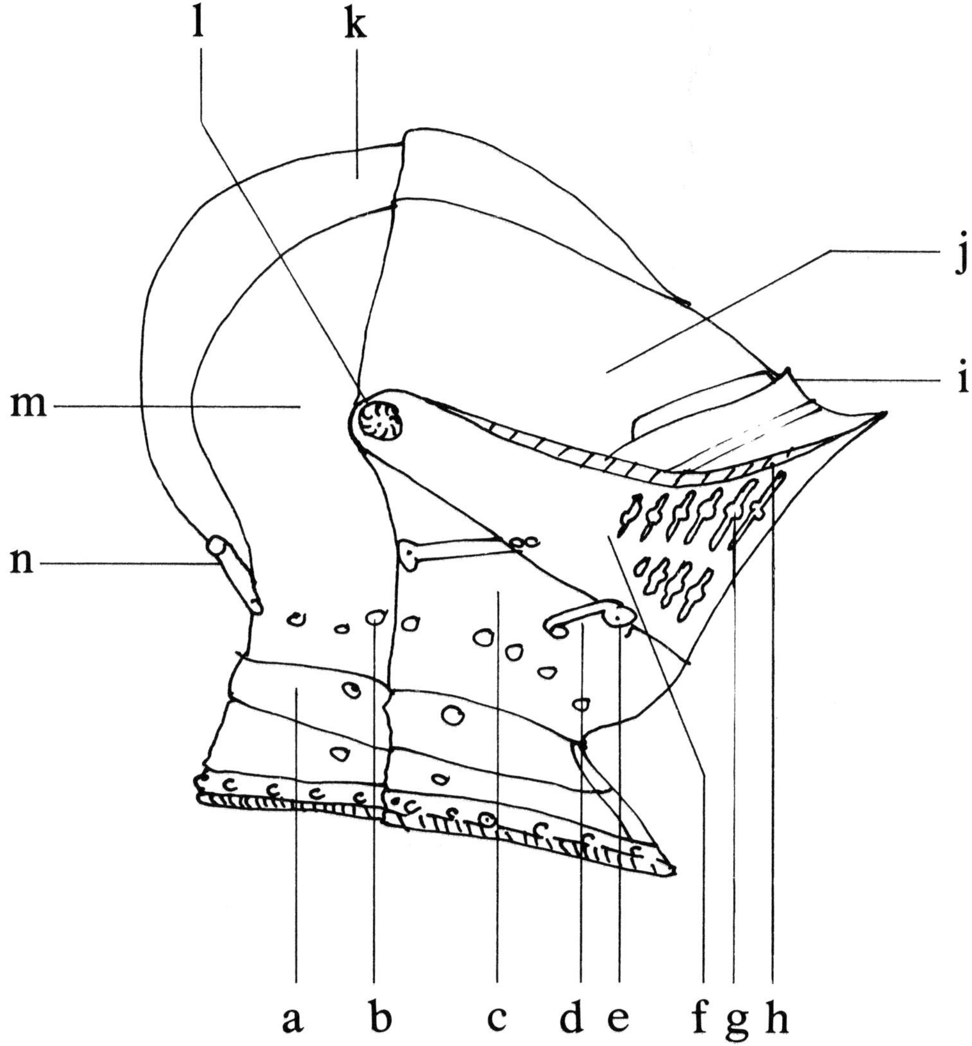
l
k
j
i
m
n
a
b
c
d
e
f
g
h

BURGONET (MORIONCINO),
ITALIAN, 3rd QUARTER OF THE 16th CENTURY.
Private Collection. Photo courtesy of Ronald A. Lee and R.T. Gwynn.

This shows some fine etching and retains some of its original gilding. The ear guards are missing.

JOUSTING SALLET (RENNHUT), GERMAN, 1575.
Musée de l'Armée, Paris (H. 52).

This fine helmet retains its original painted and engraved arabesque decoration in red, black, and gold. Note the high crest and pivoted brow reinforce, at the bottom of which are two eye slits. The threaded bolt projecting from the upper bevor was secured with straps to the cuirasse. This helmet belonged to the Prince of Radziwill; the rest of his armor is at the Kunsthistorisches Museum, Vienna. A horse armor at the Stockholm Museum with identical decoration has Nuremberg marks and the stamp of Kunz Lochner. It is probable that he made this piece.

BURGONET (MORION STURMHAUBE), GERMAN, LAST QUARTER OF 16th CENTURY, Author's Collection (244).

The skull and peak, each with a roped margin, are wrought in one piece. The neck plate, also with a roped border, is separate. The hinged cheek guards are of three lames; the lower two are restorations. Stitched to a border of leather, which in turn is riveted in place to brass capped rivets, is the original canvas quilted lining, stuffed with cow hair.

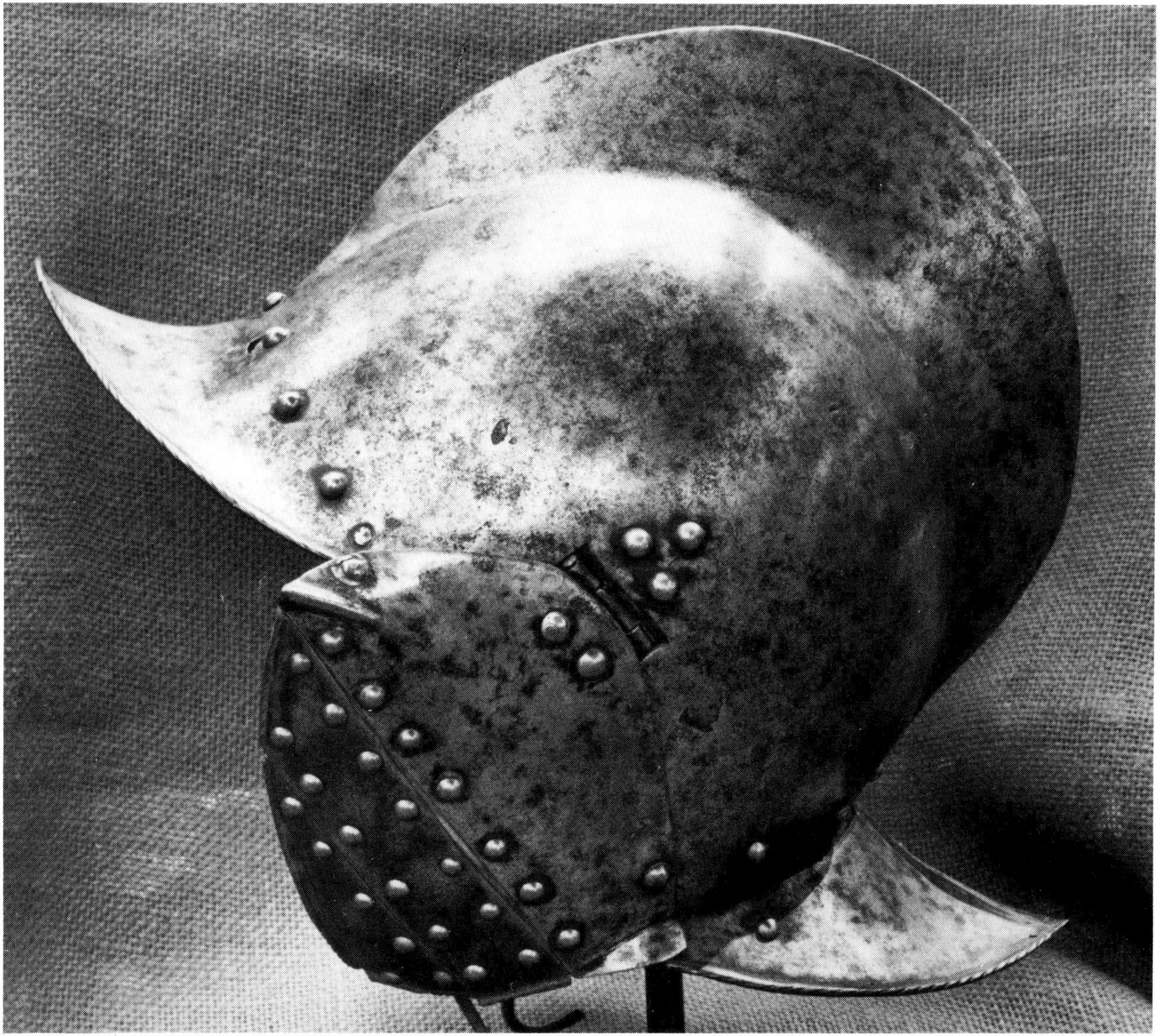

MORION CABASSET OR "SPANISH" MORION, (MORION AGUZZO, BIRNMORION), ITALIAN, c. 1575. The Metropolitan Museum of Art, New York, Bashford Dean Memorial Collection, purchase 1929. (29.158.16).

The word "birn" in German or "cabasset" in French implies a pear stem, derived from the pointed finial on helmets of this type. This one displays a fine example of etching of the period. It measures 11½ x 13½ inches.

CLOSE-HELMET, (VISIERHELM),
LOW COUNTRIES, POSSIBLY FLEMISH,
c. 1575-80.
Private Collection.

This helmet is not only beautifully made, but extremely rare because of the falling buffe, invariably a feature applied to the burgonet. The buffe consists of two lames which can be lowered a little to increase ventilation, by depressing the spring-loaded catch (visible at the lower front of the bottom buffe plate). The skull is forged from two pieces, an early date for this, and is so skillfully done it is difficult to tell in some areas. The roping on the crest and at the bottom of the gorget is all done, with superior craftmanship, by the file.

CLOSE-HELMET WITH GORGET, (ARMET À GORGERIN), FRENCH, 1580-90. Musée de l'Armée, Paris.

This represents one of the few fine quality two-piece helmets made prior to 1600. The French were pioneers in this. Note the exceptionally fine floral-mythological etching and the roped border to the edge of the stepped visor.

PARADE BURGONET, (BORGOGNOTTA), ITALIAN (BRESCIAN), LATE 16th CENTURY.

Although an excavated piece, this lovely embossed helmet depicts an artistic St. George slaying the dragon with a spear, on one side, and with a dagger on the other, and was probably worn by the guard of a high ranking Italian family. Traces of red lining remain beneath the lining rivets. The ear flaps are missing. The peak of this helmet has been strengthened (underneath) and filled in a bit, to prevent complete deterioration, as it was extremely fragile when I acquired it. A burgonet of identical shape resided in the Marzoli collection as of 1943 (See **Armi Bresciane** by Franco M. Franzo, November, 1943, pps. 43 and 44.)

MORION, GERMAN, c. 1590,
WITH SAXON MUSKETEER'S RAPIER
AND HORSEMAN'S WAR HAMMER, c. 1600.
Author's Collection (224,486,265).
All Formerly Dresden Royal Collection.

This combed morion is from the Trabanten (State) Guard of Christian I, Elector of Saxony. It is of bright steel, the surface decorated with bands of flowing foliage, etched and gilded. The center medallions contain Roman heroes: on one side, on horseback, Mutius Scaevola; and on the other, Marcus Curtius, on foot, gilded against a russeted background. On the sides of the roped comb are the insignia of the Archmarshalcy of the Holy Roman Empire and the Arms of Saxony. The plume holder is in the form of a caryatid; the holes above on each side of the comb are for securing the ostrich plume. Above the roped brim are a total of 18 lion headed rivet caps (each with a ring in its mouth) for securing the lining. The leather band to which the lining was stitched still remains. The helmet's brim is struck with the Nuremberg guild and city marks, as well as the armorer's mark, a cross with the letter ''S'' drawn through the lower half, all within a shield. These helmets were used as ''props'' in the State Opera House in Dresden as late as 1850. Morions with Fleur de Lys embossed on the sides were German, used by the Civic Guard of Munich; the emblem a symbol of the Blessed Virgin, to whom the town guard was dedicated. The comb morion saw wide use in the 16th century by shot (arquebusiers), pikemen, halbardiers, and light and heavy cavalry. Formerly Stephen V. Grancsay and Justin Stein, M.D. Collections.

A DUTCH MUSKETEER, c. 1600,
WEARING A COMB CAP **(****See page 307****).**
From Jacob de Gheyn,
The Exercise of Armes for Calivres, Muskettes, and Pikes,
the Hague 1608.

IDGheyn. in.

CABASSET (SPANISH MORION, ZUCOTTO), ITALIAN, c. 1600.
Author's Collection (209).

Engraved with acanthus leaves over the medial ridge (front and back) and at the apex on either side, and with "U's" and small arrows around the brim, this deeply russeted helmet has a plume holder and brass petals decorating the rivet caps for the lining. This type of helmet was derived from the Spanish cabacete and was a favorite headpiece of the musketeer, as it did not interfere with his aiming. The short stalk (cabasset means "pear stem"), narrow brim, and almond shaped skull are characteristic of the cabasset.

**SAVOYARD HELMET,
(DEATH'S HEAD BURGONET,
ELMETTO ALLA SAVOIARDA, TODENKOPF),
ITALIAN, c. 1600.
Frank Andrina Collection.**

This helmet represents a distinctive type of siege weight helmets used by field commanders when attacking a town wall or fortress. Its unique characteristics are the hinged "V"-shaped fall, or brow reinforce, with its projecting eye shades; circular holes pierced for the eyes; and holes pierced over the nose and mouth, vaguely suggestive of a skull. Many were captured from the Savoyards in their attack on Geneva on December 12, 1602, in which Duke Charles Emmanuel I, disregarding previous treaties, made a final attempt to take Geneva by surprise by scaling the walls with ladders. The "night of the escalade" resulted in bitter defeat for the House of Savoy, which had been attempting to capture Geneva since the year 1290.

COMB CAP (COMBED HEAD-PIECE, POT), POSSIBLY DUTCH, c. 1620.
Author's Collection (177).

One of the most common types of head piece (2 piece construction) in the early 17th century, it is now seldom seen in collections, probably because it didn't seem valuable enough to save. This was the type worn by pikemen and musketeers of the 1600-1620 period and by caliver men to 1630. It has been painted black to reduce glare and prevent rust. Note the three rivets at the center above the narrow brim for the ear-flap strap mount. Some continental examples seem to be a bit higher and have a more pronounced step at the apex.

CLOSE-HELMET WITH GORGET, (ARMET À GORGERIN), FRENCH, 1600-25. Musée de l'Armée, Paris.

The skull has two piece construction, which became common in the 17th century. It has a high rolled comb with acorn finial and a plume holder at the back. The decoration consists of unusually fine etching and retains traces of original gilding.

CUIRASSIER HELMET (HORSEMAN'S CLOSE HELMET), ENGLISH, c. 1625.
Author's Collection (249).

It has a two-piece skull with a moderate comb; a pivoted peak and three barred visor (moving as one unit). The bevor and front gorget plate swing forward from the skull and rear gorget plate by releasing the "s" hook for donning. This example retains a red velvet lining and black surface paint. The cuirassier head piece developed from the burgonet and first appeared about 1600.

"DUTCH" POT HELMET
(ZISCHÄGGE, LOBSTER TAILED POT),
HUNGARIAN OR GERMAN, c. 1630.
Private Collection.

Many of this type were imported into England for use in the Civil War (1642-49). It has a one piece ribbed skull terminating in a ring shaped finial. Each of the four neck lames and the peak is punched with four series and three dots. The peak is also marked with a "B," and the rear neck lame is struck with "ZH3." A heart shaped screw bolt holds the adjustable sliding nasal in the position desired. These helmets were most popular with horsemen from c. 1640, some Eastern European examples having a much longer neckguard extending to the back (see example page 323).

SPIDER HELMET (SIEGE HELMET), FRENCH, c. 1640. Author's Collection (260).

A fascinating headpiece constructed of light, very hard high quality steel, this weighs only 3 lbs. but affords as much protection when climbing a ladder or fighting beneath the walls of a fortress, as some of the 15 lb. types of the period. When the lever (front center between peak and finial) is depressed, the circular 4 inch diameter finial plate springs upward allowing the 9 hinged bar guards to be locked upwards in position against the vertical framework of the skull piece. The wide hinged neck guard remains down. The spider helm is fairly rare, examples found at the Metropolitan Museum of Art, Tower of London, and the Porte de Hal, Brussels. This one measures 7½ inches high; 12½ inches with the bar guards down.

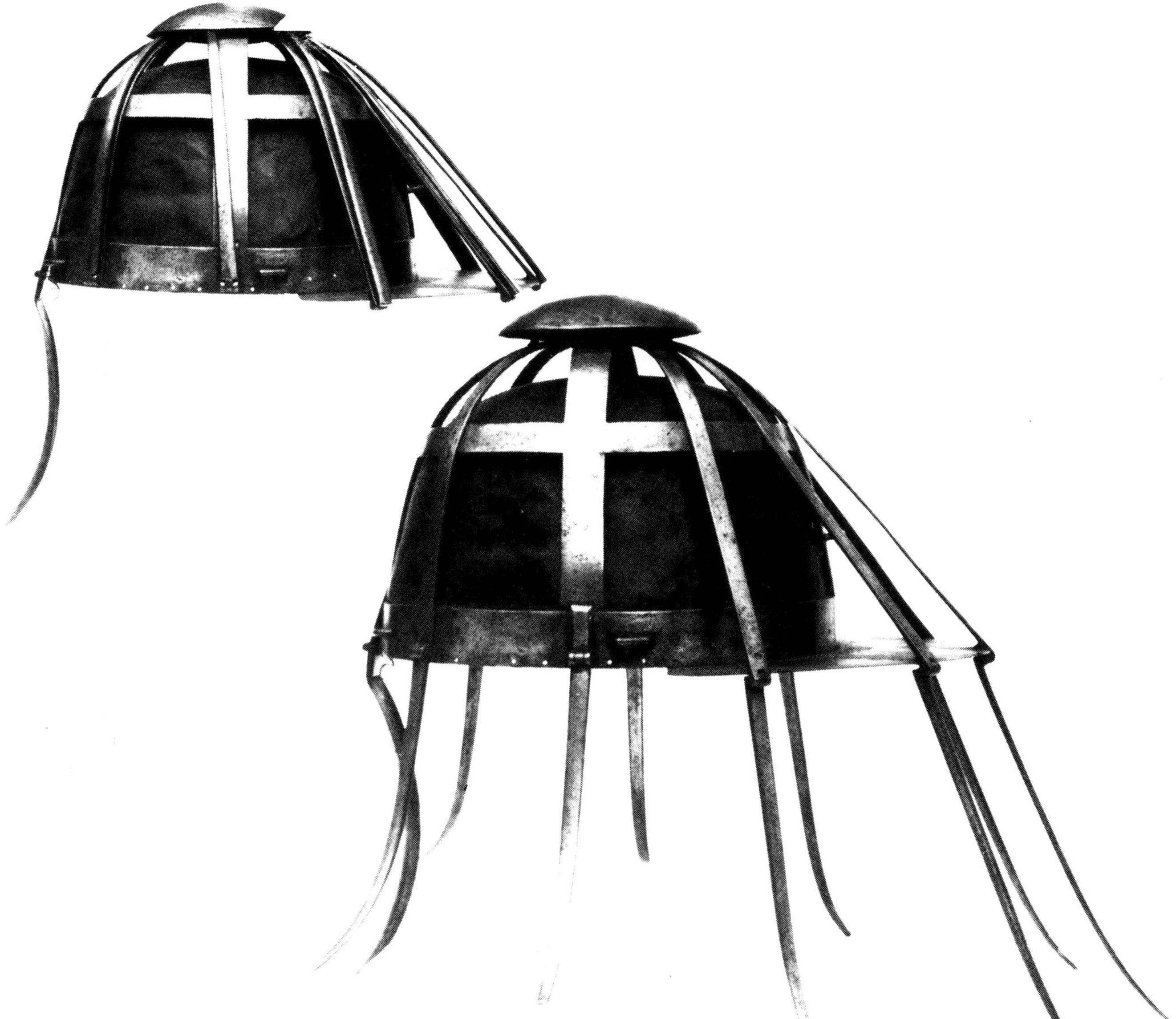

PIKEMAN'S POT HELMET, ENGLISH, 1620-50.
Private Collection.

This represents the most popular helmet of the English pikeman of this period, with two-piece skull, moderate comb, and wide, downward sloping brim, curving to a peak fore and aft. This type is also called a comb cap or pot. All versions of pot helmets had small tapering cheekpieces which tied under the chin, and usually a leather loop was riveted under the rear of the brim for hanging on a hook on the backplate when it was not in use. Sometimes a plume holder is evident also. The crowned ''A'' of the London Armourer's Company is punched on the left front brim, flanked by the initials ''M'' and ''V.'' The London Armorer's Company, founded in the 14th century, changed its mark by adding a helmet (instead of the crown) over the ''A'' from 1649-1660 (Commonwealth).

STEEL HAT (CHAPEL D'ACIER),
FRENCH, LOUIS XIII (1st HALF OF 17th CENTURY),
Musée de l'Armée, Paris.

The helmet and gorget
(this type was made only
for wear with a buff coat)
retain most of their
original blue and gilded finish.

SIEGE HELMET, SWISS, c. 1650.
Author's Collection (179).

An Armory number (consisting of 5 dots) is punched at the base of the one piece skull just above the peak at the edge of the narrow comb. An indistinguishable maker's mark is struck at the "V" of the peak. (The ear flaps are missing.) This is the type of shot-proof helmet used by field commanders for observation or when directing siege operations. Weight 9 lbs.

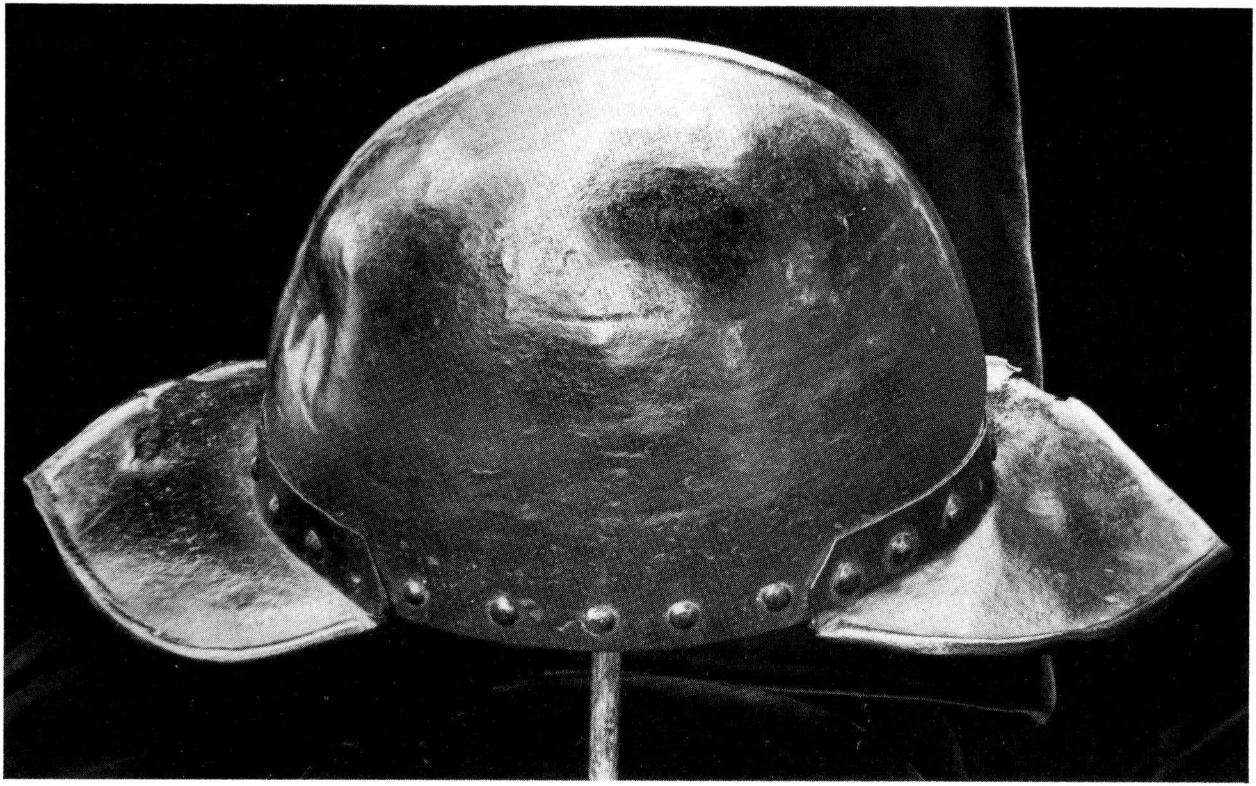

ZISCHÄGGE (LOBSTER-TAILED POT,
LOBSTER-TAILED BURGONET),
EAST GERMAN, 2nd QUARTER OF 17th CENTURY.
Glasgow Art Gallery and Museum, Kelvingrove, Glasgow.

This is a fine example of the favorite light horseman's helmet from 1620 on. The laminated neckguard and pivoted (or stationary) peak (fall) and sliding nasal (here missing) are typical. The fluted skull, finial, and fine floral etching add to the quality of this one. This type of helmet was made by German armorers in similar style to the Turkish helmets used in Eastern Europe. The English version, used in the Civil War (1642-49) usually had a low comb on the skull, a triple barred face guard attached underneath the peak (instead of a nasal bar), and a (solid) neck guard with simulated lames on the cheaper versions (see page 337). After 1650 the skull usually had no comb at all.

OFFICER'S ZISCHARGGE, (POT, LOBSTER-TAILED BURGONET, STURMHAUBE), GERMAN OR FLEMISH EARLY SEVENTEENTH CENTURY.

Many variations of this style helmet were in use throughout the continent and in England in the seventeenth century, as it was the type favored by cavalrymen. The skull is deeply ribbed, terminating in a rectangular finial. The sliding, adjustable nasal terminates in a pierced spatulate design. The neck guard consists of five lames, surmounted by a tapered plume holder. This helmet is complete with its original silk lining, padded with wool, and shows the influence of Hungarian styling of the second half of the sixteenth century.

OFFICER'S ZISCHAGGE, (POT, LOBSTER-TAILED BURGONET, STURMHAUBE), GERMAN, POLISH OR FLEMISH, EARLY SEVENTEENTH CENTURY.

Front view.

BURGONET, (BOURGUIGNOTTE A VENTAILLE), Musee de l'Armee, Paris, 17th Century.

This helmet, with its two-piece skull, shows great quality with its pleated skull and fine etching. The cheek pieces open outward like an armet on hinges at each lower side of the skull. It belonged to the Infant Dauphin, Louis XIV.

IRON HAT, ENGLISH,
2nd QUARTER OF 17th century,
Glasgow Art Gallery and Museum, Kelvingrove, Glasgow.

This rare type, a copy in steel of the broad-brimmed civilian ''cavalier's'' hat, was used by horsemen during the civil war period. Note the plume holder and adjustable wing nut for a sliding bar nasal guard (missing).

POT HELMET WITH HARQUEBUS ARMOR OF JAMES II, ENGLISH, 1686.
Tower of London Armouries (II.123).

This was made in London by Richard Hoden, and is displayed with a left elbow gauntlet and buff leather coat. The royal arms and initials IR (Iacobus Rex) appear on the helmet's face guard. A similar type, but of more elaborate workmanship to the pot helm with triple barred face guard, shown on p. 337.

PATE DEFENSE, (SECRETE), CONTINENTAL OR ENGLISH, 17th CENTURY. The Metropolitan Museum of Art, New York. Gift of William Henry Riggs, 1914.

This skull cap formed of narrow bands of steel riveted into a cage-like structure measures 4 inches high by 7-5/8 inches in length. It was originally lined with canvas stuffed with tow and inserted into a civilian cavalier's hat, allowing the protection of a helmet without giving that appearance, and because of its construction remaining extremely light.

POT HELMET, ENGLISH,
2nd HALF OF 17th CENTURY.
Private Collection.

The faceguard, which is riveted to the hinged peak, consists of three vertical bars joined together at the bottom in a point. The neck guard is of one piece, simulating three lames (on less expensive models). And the heavy two-piece skull overlaps into a very low comb, and is joined by six rivets. The ear flaps are missing. One siege style of this helmet had curved supports (suspended from vertical bars attached to the sides of the skull) which rested on the wearer's shoulders. This is the pure English version of the Continental lobster-tailed burgonets.

SAPPER'S HELMET, (SAPPENHELM), GERMAN, 2nd HALF OF 17th CENTURY.
Author's Collection (257), Formerly William C. Borland Collection.

This heavily constructed 14 lb. helmet has a two-piece skull with a raised crest or comb. The hinged ear flaps, curved peak and hinged neck guard are restorations by R.W. MacWillie. Large brass rivets are used. Helmets of this type were used as late as 1848, represented in the ''Siege of Rome'' by Raffet, and may well have been ''leftovers'' from the 17th century. Similar helmets may be seen in the Tower of London and in the Museum Für Deutsche Geschichte, Berlin. The ear guards, peak, and neck guards were frequently lost on this type of helmet (they were of lighter construction than the extremely thick skull), giving the skull piece a modified ''barbuta'' look, confusing one who studies it without being aware of its original configuration.

HORSEMAN'S ARMOR,
(CORSALETTO DA CAVALLO LEGGERO),
BRESCIAN, c. 1680-90.
Musee de l'Armee, Paris.

Accompanied by a matching chanfron, this armor with its attractive original blue "stripes," is signed by its maker, Lorenzo Saiano. Note the unusual shape of the ear flaps on the helmet and the expanded laminated culet of the harness, giving added protection as the wearer sits in the saddle. Movable gussets have disappeared, as has the fauld. The laminated tassets are attached to the flange on the breastplate. Here we see an example of unusually fine workmanship from the end of the period of decline in armor, brought about by the improvements in gunpowder.

VENETIAN PARADE BARBUT
(BARBUTA ALLA VENEZIANA), 18th CENTURY?
Odescalchi Collection, Rome.

The decoration on this ceremonial barbut was done in the 17th or 18th century. The crest finial is missing. It is difficult to tell without dismantling and possibly damaging it whether or not the helmet beneath is actually of 15th century manufacture, as many of these decorated helmets were. A good example is a similar helmet, damaged in the Bargello Museum from floodwaters in 1966, which was dismantled and restored along with many other fine objects by a group of specialists at the Kunsthistorisches Museum in Vienna.

1092

BIBLIOGRAPHY

Blair, C., EUROPEAN ARMOUR, London, 1958.

Boccia, L.G. and Coelho, E.T., L'ARTE DELL'ARMATURA IN ITALIA, Milan 1967.

Boheim, W., ALBUM AUS DER WAFFENSAMMLUNG DES ALLERHOCHSTEN KAISER-HAUSES, Wien, 1894. Vol. I, II.

Clephan, R.C., THE TOURNAMENT, ITS PERIODS AND PHASES, London, 1919.

D'Anjou, R., LE LIVRE DES TOURNOI DU ROI RENE, Paris, 1946.

De Cosson, Baron, and Burgess, W., ANCIENT HELMETS AND EXAMPLES OF MAIL, London, 1881.

Demmin, A., AN ILLUSTRATED HISTORY OF ARMS AND ARMOUR, London, 1911.

De Riquer, M., L'ARNES DELL CAVALLER, Barcelona, 1968.

Dufty, A.R., EUROPEAN ARMOUR IN THE TOWER OF LONDON, London, 1968.

Ffoulkes, C., INVENTORY AND SURVEY OF THE ARMOURIES OF THE TOWER OF LONDON, London, 1916, Vol. I, II.

Gamber, O., GLOSSARIUM ARMORUM, ARMA DEFENSIVA, Graz, 1972.

Gimbel, K., GIMBEL COLLECTION AUCTION CATALOGUE, Berlin, 1904.

Grancsay, S.V., "The New Galleries of European Arms and Armor," THE METROPOLITAN MUSEUM OF ART BULLETIN, New York, May 1956, p. 205.

——— "Sculpture in Steel," THE METROPOLITAN MUSEUM OF ART BULLETIN, New York, January, 1963, p. 182.

Harmand, A., JEANNE D'ARC, SES COSTUMES, SON ARMURE, Paris, 1929.

Hencken, H., THE EARLIEST EUROPEAN HELMETS, Cambridge, 1971.

Hewitt, J., ANCIENT ARMOUR, London, 1855. Vols. I, II, III.

Kelly, F.M. and Schwabe, R., A SHORT HISTORY OF COSTUME AND ARMOUR, 1066-1800, London, 1931.

Von Kienbusch, C.O. and Grancsay, S.V., THE BASHFORD DEAN COLLECTION OF ARMS AND ARMOUR, Portland, 1933.

Laking, Sir G.F., A RECOND OF EUROPEAN ARMS AND ARMOUR THROUGH SEVEN CENTURIES, London, 1922. Vols. I-V.

Mann, Sir J.G., THE ARMOURY OF THE CASTLE OF CHURBURG, London, 1929.

——— WALLACE COLLECTION CATALOGUE, London, 1962, Vol. I.

Martin, P., ARMS AND ARMOUR, Rutland, 1968.

Muller, H., and Kunter, F., EUROPAISCHE HELME, Berlin, 1971.

Niox, General, MUSEE DE L'ARMEE: ARMES AND ARMURES ANCIENNES, Paris, 1917. Vol. I.

Oakeshott, R.E., THE ARCHAEOLOGY OF WEAPONS, London, 1960.

——— A KNIGHT AND HIS ARMOUR, London, 1965.

Robinson, H.R., THE ARMOUR OF IMPERIAL ROME, London, 1975.

Rossi, F., and Di Carpegna, N., ARMI ANTICHE, Milano, 1969.

Schneider, Hugo, ''Die beiden Topfhelme von Madeln,'' ZEITSCHRIFT FUR SCHWEIZERISCHE ARCHAOLOGIE UND KUNSTGESCHICHTE, Vol. 14, 1953, pp. 24-46.

Scott, R.L., THE SCOTT COLLECTION OF ARMS AND ARMOUR, Glasgow, 1924, Vol. I.

Snodgrass, A.M., ARMS AND ARMOUR OF THE GREEKS, London, 1967.

Stenten, Sir F., THE BAYEUX TAPESTRY, London, 1965. (Chapter on ''Arms and Armour'' by Sir James Gow Mann.)

Terenzi, M., ELMI DEL TRECENTO ALLA MOSTRA DEI POPPI, Rome, 1967.

———MOSTRA DI ARMI ANTICHE, Poppi Castello De Conti guidi, Firenze, 1967.

Wegeli, R., INVENTAR DER WAFFENSAMMLUNG DES BERNISCHEN HISTORISCHEN MUSEUMS IN BERN, Bern, 1939. Vol. I.